SpringerBriefs in Mathematics

SpringerBriefs present concise summaries of cutting-edge research and practical applications across a wide spectrum of fields. Featuring compact volumes of 50 to 125 pages, the series covers a range of content from professional to academic. Briefs are characterized by fast, global electronic dissemination, standard publishing contracts, standardized manuscript preparation and formatting guidelines, and expedited production schedules.

Typical topics might include:

- A timely report of state-of-the art techniques
- A bridge between new research results, as published in journal articles, and a contextual literature review
- A snapshot of a hot or emerging topic
- An in-depth case study
- A presentation of core concepts that students must understand in order to make independent contributions

SpringerBriefs in Mathematics showcases expositions in all areas of mathematics and applied mathematics. Manuscripts presenting new results or a single new result in a classical field, new field, or an emerging topic, applications, or bridges between new results and already published works, are encouraged. The series is intended for mathematicians and applied mathematicians. All works are peer-reviewed to meet the highest standards of scientific literature.

Titles from this series are indexed by Scopus, Web of Science, Mathematical Reviews, and zbMATH.

Giuseppe Maria Coclite

Scalar Conservation Laws

 Springer

Giuseppe Maria Coclite
Department of Mechanics, Mathematics
and Management
Polytechnic University of Bari
Bari, Italy

ISSN 2191-8198   ISSN 2191-8201   (electronic)
SpringerBriefs in Mathematics
ISBN 978-981-97-3983-7   ISBN 978-981-97-3984-4   (eBook)
https://doi.org/10.1007/978-981-97-3984-4

Mathematics Subject Classification: 35L60, 35L65

This Springer imprint is published by the registered company Springer Nature Singapore Pte Ltd.
The registered company address is: 152 Beach Road, #21-01/04 Gateway East, Singapore 189721,
Singapore

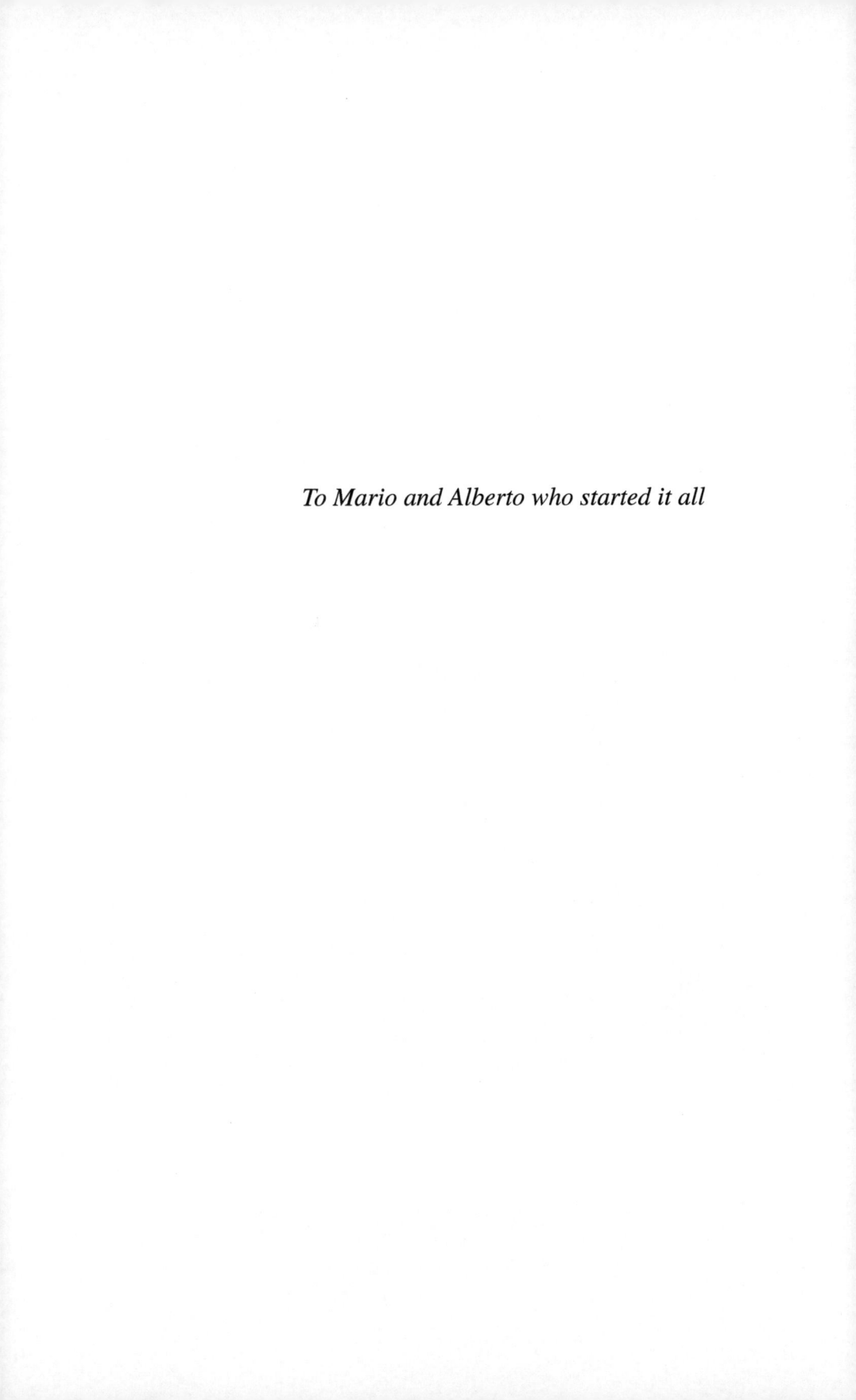

To Mario and Alberto who started it all

Preface

These notes provide a self-contained introduction to the mathematical theory of hyperbolic scalar conservation laws in one space dimension, which are first-order partial differential equation of the type

$$\partial_t u + \partial_x f(u) = 0, \tag{1}$$

where u is the *conserved quantity*, and f is the *flux (of the conserved quantity)*.

One of the features of (1) is the spontaneous creation of singularities: a shock can appear during the evolution of the equation independently on the smoothness of the initial datum. As a consequence, the solutions need to be interpreted in a weak sense and found in spaces of discontinuous functions.

Hyperbolic conservation laws are fundamental in continuum physics; they appear naturally in modelling the dynamics of gases and fluids (see the classical Euler and Burgers equations). Since the middle of the last century, after the famous papers of Lighthill and Whitham [4] and Richards [5] hyperbolic conservation laws have been used for the description of vehicular dynamics, indeed the spontaneous creation of singularities models the occurrence of traffic jams without bottlenecks, which is one of the peculiar features of vehicular traffic.

Chapter 1 is the introduction and provides qualitative answers to three questions:

1. Why do we use the terms *conservation law, conserved quantity*, and *flux*?
2. Which kind of physical phenomena is an hyperbolic conservation law able to describe?
3. Which are the mathematical features of the solutions of hyperbolic conservation laws?

In particular, we deduce the Burgers equation and the Lighthill, Whitham, and Richards traffic model.

Chapter 2 is dedicated to the method of characteristics. We explicitly solve first order equations of the type

$$\partial_t u + a(t, x)\partial_x u = h(t, x, u).$$

Moreover, we show that if the coefficient a depends on the unknown u, that is

$$\partial_t u + a(t, x, u)\partial_x u = h(t, x, u),$$

the catastrophe of the gradient may occur, namely characteristics cross, and there is the creation of discontinuities in finite time even if the initial datum is analytic.

Chapter 3 is devoted to the concept of weak entropy solutions. We give the definition of distributional solution and prove that a shock is a distributional solution if and only if the Rankine-Hugoniot Condition holds. We then provide the definition of entropy solution and prove a necessary and sufficient condition for a shock to be an entropy solution. We conclude the chapter with the Kružkov theorem stating the uniqueness and stability of the entropy solution to a Cauchy problem.

Chapter 4 is dedicated to the explicit construction of entropy solution of the Riemann problem, that is a Cauchy problem with an initial datum of Heaviside type.

In Chap. 5, we review the space of functions with bounded variation in one and two dimensions. Indeed, BV is the space in which entropy solutions live if the initial data belong to BV. We give several examples and prove some qualitative result and compactness theorems that play a key role in the existence results. The proofs and the definitions require only the knowledge of some basic measure theory because we do not consider the general n-dimensional case.

In Chaps. 6, 7, and 8 we present three different approaches to the existence issue: the wave front-tracking, the vanishing viscosity, and the compensated compactness. Front-tracking consists in the approximation of conservations laws with equations of the same type but with piecewise affine fluxes and piecewise constant initial data. It has been extremely successful in the analysis of systems [1] and control problems [3]. Vanishing viscosity is based on a parabolic approximation of the hyperbolic equation. It is interesting for the reminiscence of the convergence of the Naiver-Stokes equations towards the Euler ones in fluid-dynamics. Moreover, it gives precious hints on the convergence analysis of numerical schemes [3]. Finally, compensated compactness is a useful measure theoretic tool dealing with the compression effects of nonlinear equations. It works in the nonlinear case and applies to several convergence problems. Indeed in Chap. 9, we use compensated compactness in order to study the asymptotic decay of periodic solution.

Chapter 10 is dedicated to the Oleinik estimate, that is, a one-side Lipschitz inequality, that holds if the flux is strictly convex and is equivalent to the entropy inequalities.

Finally, in Chap. 11, we present the Lax-Oleinik Formula that provides an explicit formula for the solutions of the initial value problem when the flux is strictly convex.

These notes have developed from the lecture notes prepared for several PhD courses held in the Università di Bari, the Politecnico di Bari, the Politecnico di Torino, and a CIME summer course [2]. They can be used for courses at graduate level. Indeed, they provide the basic and classical results on hyperbolic conservations laws and all the technical details are provided. For the same reasons, they can be useful also for researchers.

Bari, Italy   Giuseppe Maria Coclite

References

1. Bressan, A.: Hyperbolic Systems of Conservation Laws – The One-Dimensional Cauchy Problem. Oxford Lecture Series in Mathematics and its Applications, vol. 20. Oxford University Press, Oxford (2000)
2. Coclite, G.M., Maddalena, F.: Conservation laws in continuum mechanics. In: Applied Mathematical Problems in Geophysics, Cetraro, Italy, July 2019. Lecture Notes Given at the Summer School, pp. 157–207. Springer/Fondazione CIME, Cham/Florence (2022)
3. Dafermos, C.M.: Hyperbolic Conservation Laws in Continuum Physics. Grundlehren der Mathematischen Wissenschaften [Fundamental Principles of Mathematical Sciences], vol. 325, 4th edn. Springer-Verlag, Berlin (2016)
4. Lighthill, M.J., Whitham, G.B.: On kinematic waves. I. Flood movement in long rivers. Proc. R. Soc. Lond. Ser. A. **229**, 281–316 (1955)
5. Richards, P.I.: Shock waves on the highway. Oper. Res. **4**, 42–51 (1956)

Acknowledgements

I warmly thank my friends and colleagues Professors Alessandro Coclite, Nicola De Nitti, Carlotta Donatello, Francesco Maddalena, and Nils Henrik Risebro for carefully reading the manuscript and their many useful suggestions, comments, and remarks.

Contents

Chapter 1
Introduction

Abstract This chapter provides qualitative answers to three questions:

(1) Why do we use the terms *conservation law, conserved quantity*, and *flux*?
(2) Which kind of physical phenomena is an hyperbolic conservation law able to describe?
(3) Which are the mathematical features of the solutions of hyperbolic conservation laws?

In particular, there is the deduction of the Burgers equation and the Lighthill, Whitham, and Richards traffic model.

Keywords Conserved quantity · Traffic models · Lighthill-Whitham-Richards model · Burgers equation · Euler equations · Aw-Rascle model

These notes are devoted to the study of the first order partial differential equation

$$\partial_t u + \partial_x f(u) = 0, \tag{1.1}$$

where $f \in C^2(\mathbb{R}^N; \mathbb{R}^N)$, $u : [0, \infty) \times \mathbb{R} \to \mathbb{R}^N$, and $N \geq 1$. The function $u = u(t, x)$ is termed *conserved quantity*, $f = f(u)$ *flux (of the conserved quantity)*. If $N = 1$ we say that (1.1) is a *scalar conservation law*, if $N > 1$ we say that (1.1) is a *system of conservation laws* and it stands for

$$\begin{cases} \partial_t u_1 + \partial_x f_1(u_1, \dots, u_N) = 0, \\ \dots \dots \\ \partial_t u_N + \partial_x f_N(u_1, \dots, u_N) = 0, \end{cases}$$

where

$$u = u(t, x) = (u_1(t, x), \dots, u_N(t, x)),$$
$$f = f(u) = (f_1(u_1, \dots, u_N), \dots, f_N(u_1, \dots, u_N)).$$

Fig. 1.1 Flow trough the end points

In this section we try to answer to the following questions:

(1) Why do we use the terms *conservation law*, *conserved quantity*, and *flux* for (1.1), u, and f, respectively?
(2) Which kind of physical phenomena is (1.1) able to describe?
(3) Which are the mathematical features of the solutions of (1.1)?

Let us answer to the first question. If u is a smooth solution of (1.1) and $a < b$ we have that (see Fig. 1.1)

$$\frac{d}{dt} \int_a^b u(t,x)dx = \int_a^b \partial_t u(t,x)dx$$

$$= - \int_a^b \partial_x f(u(t,x))dx$$

$$= f(u(t,a)) - f(u(t,b))$$

$$= [\text{inflow at } x = a \text{ and time } t]$$

$$- [\text{outflow at } x = b \text{ and time } t].$$

In other words, the conserved quantity u is neither created nor destroyed, the amount of u in the interval $[a, b]$ changes only depending on the flow through the two end-points.

We answer to the second question showing some physical models expressed in terms of conservation laws. We begin with the road fluid-dynamic traffic model introduced by Lighthill, Whitham, and Richards [2, 3]. Consider a one way one lane infinite road and let $\rho = \rho(t, x)$ be the the density of vehicles at time t in the position x. Assuming that the vehicles behave as fluid particles we have

$$\partial_t \rho + \partial_x (\rho v) = 0, \tag{1.2}$$

where v is the velocity of the vehicles. The key assumption of Lighthill, Whitham, and Richards is that the velocity depends only on the density, namely

$$v = v(\rho), \tag{1.3}$$

that is somehow reasonable in case of highways. The drivers regulate their velocity in function of the number of vehicles in front of them. Therefore writing

$$f(\rho) = \rho v(\rho),$$

(1.2) reads

$$\partial_t \rho + \partial_x f(\rho) = 0. \tag{1.4}$$

On $v = v(\rho)$ it is reasonable to assume that

$$v(0) = v_{max}, \qquad v(\rho_{max}) = 0, \qquad v \text{ is decreasing,}$$

namely more vehicles are on the road slower they go and viceversa less vehicles are on the road faster they go. In particular, Lighthill, Whitham, and Richards proposed

$$v(\rho) = v_{max} \left(1 - \frac{\rho}{\rho_{max}} \right).$$

We can deduce one more simple model considering a one-dimensional material made of non interacting particles, for example a low dense gas. More precisely, our deduction is valid as soon as the particles do not interact and their paths do not cross. We can identify the particles using their initial position y. Let $\varphi(t, y)$ be the position at time t of the particle that at time $t = 0$ was in y, its velocity and acceleration are $\partial_t \varphi$ and $\partial_{tt}^2 \varphi$, respectively. Since the particles do not interact within themselves, we cannot have two different particles in the same position at the same time, therefore $\varphi(t, \cdot)$ is increasing and, in particular, invertible. Let $\psi(t, \cdot)$ be the inverse of $\varphi(t, \cdot)$, i.e.,

$$y = \psi(t, \varphi(t, y))$$

and

$$x = \varphi(t, y) \iff y = \psi(t, x).$$

Let $u(t, x)$ be the velocity of the particle that at time t is in x, namely

$$x = \varphi(t, y),$$
$$u(t, x) = u(t, \varphi(t, y)) = \partial_t \varphi(t, y),$$
$$u(t, x) = \partial_t \varphi(t, \psi(t, x)).$$

The acceleration of the particle that at time t is in x is

$$\partial_{tt}^2 \varphi(t, y) = \partial_t \left(\partial_t \varphi(t, y) \right) = \partial_t \left(u(t, \varphi(t, y)) \right)$$
$$= \partial_t u(t, \varphi(t, y)) + \partial_x u(t, \varphi(t, y)) \partial_t \varphi(t, y)$$
$$= \partial_t u(t, x) + \partial_x u(t, x) u(t, x).$$

Since the particles do not interact among themselves, there are no forces acting on them. Then, due to the Second Law of Dynamics, the acceleration is zero. As a consequence the previous equation becomes

$$\partial_t u + \partial_x \left(\frac{u^2}{2} \right) = 0, \tag{1.5}$$

that is termed *Burgers equation*.

The Lighthill-Whitham-Richards traffic model and the Burgers equation are models expressed in terms of scalar conservation laws, we continue with two more models expressed in terms of systems of conservation laws.

The Euler equations for a non-viscous compressible gas in Lagrangian coordinates are

$$\begin{cases} \partial_t v - \partial_x u = 0, & \text{(conservation of mass)} \\ \partial_t u + \partial_x p = 0, & \text{(conservation of momentum)} \\ \partial_t \left(e + \frac{u^2}{2} \right) + \partial_x (up) = 0, & \text{(conservation of energy)} \end{cases} \tag{1.6}$$

where v is the specific volume (i.e., $1/v$ is the density), u is the velocity, e is the energy, and p is the pressure of the gas. Since we have three equations in four unknowns, we need the constitutive equation of state

$$p = p(e, v),$$

that depends on the specific gas under consideration.

Finally, we have the traffic model proposed by Aw and Rascle [1]

$$\begin{cases} \partial_t \rho + \partial_x \left(y + \rho^{\gamma+1} \right) = 0, \\ \partial_t y + \partial_x \left(\frac{y^2}{2} - y\rho^\gamma \right) = 0, \end{cases} \tag{1.7}$$

where ρ is the density, y this the generalized momentum of the vehicles, and γ is a positive constant. It is an extension of the Lighthill, Whitham, and Richards model

$$\partial_t \rho + \partial_x (\rho v(\rho)) = 0.$$

Indeed in (1.7) we have

$$\partial_t \rho + \partial_x (\rho v(y, \rho)) = 0, \qquad v(y, \rho) = \frac{y}{\rho} + \rho^\gamma,$$

where the velocity v depends on the density ρ and the generalized momentum y.

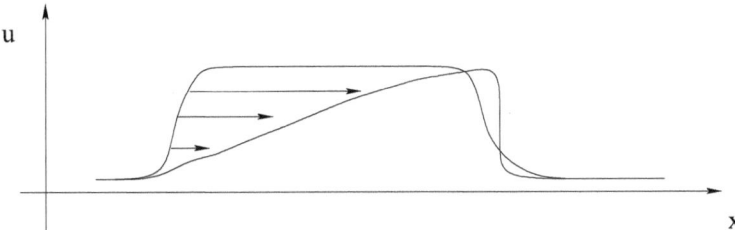

Fig. 1.2 Spontaneous creation of a shock

Regarding the third question, one of the main features of conservation laws relies in solutions which exhibit the creation of discontinuities. Indeed, even scalar problems with analytic flux and initial condition, like

$$\begin{cases} \partial_t + \partial_x \left(\dfrac{u^2}{2} \right) = 0, & t > 0, \ x \in \mathbb{R}, \\ u(0, x) = \dfrac{1}{1 + x^2}, & x \in \mathbb{R}, \end{cases} \tag{1.8}$$

experience the creation of discontinuities in finite time , see Fig. 1.2.

A detailed analysis of (1.8) can be found in Sect. 2.3.

References

1. Aw, A., Rascle, M.: Resurrection of "second order" models of traffic flow. SIAM J. Appl. Math. **60**(3), 916–938 (2000)
2. Lighthill, M.J., Whitham, G.B.: On kinematic waves. I. Flood movement in long rivers. Proc. R. Soc. Lond. Ser. A. **229**, 281–316 (1955)
3. Richards, P.I.: Shock waves on the highway. Oper. Res. **4**, 42–51 (1956)

Chapter 2
The Method of Characteristics

Abstract This chapter is dedicated to the method of characteristics. We explicitly solve first order equations of the type

$$\partial_t u + a(t, x)\partial_x u = h(t, x, u).$$

Moreover, we show that if the coefficient a depends on the unknown u

$$\partial_t u + a(t, x, u)\partial_x u = h(t, x, u)$$

the catastrophe of the gradient may occur, namely characteristics cross and there is the creation of discontinuities in finite time even if the initial datum is analytic.

Keywords Characteristics · Catastrophe of the gradient

Before starting our analysis on hyperbolic conservation laws it is useful to review the classical method of characteristics used for hyperbolic equations of the type

$$\partial_t u + a(t, x)\partial_x u = h(t, x, u), \tag{2.1}$$

that may a have a nonlinear dependence on u only in the source term $h(t, x, u)$. The method of the characteristics gives a complete description of the solution of (2.1). Unfortunately it does not produce global in time solutions in the general case when a depends also on u

$$\partial_t u + a(t, x, u)\partial_x u = h(t, x, u),$$

due to the crossing of the characteristics and the creation of discontinuities.

The method of characteristics for equation in more than one spacial dimension is treated in [1, Chapter 3].

G. M. Coclite, *Scalar Conservation Laws*, SpringerBriefs in Mathematics,
https://doi.org/10.1007/978-981-97-3984-4_2

2.1   Hyperbolic Equations with Constant Coefficients

In order to make our presentation as clear as possible we start with the simplest possible case in which the coefficient a is constant in time and space.

Consider the Cauchy problem

$$\begin{cases} \partial_t u + a\partial_x u = h(t, x, u), & t > 0, \ x \in \mathbb{R}, \\ u(0, x) = u_0(x), & x \in \mathbb{R}, \end{cases} \tag{2.2}$$

where $u_0 \in C^1(\mathbb{R})$, $h \in C^1([0, \infty) \times \mathbb{R}^2)$, and $a \in \mathbb{R}$ is a constant.

Define

$$v = \begin{pmatrix} 1 \\ a \end{pmatrix}.$$

The method of the characteristics is based on the following observation

$$\partial_v u = (\partial_t u, \partial_x u) \cdot \begin{pmatrix} 1 \\ a \end{pmatrix} = \partial_t u + a\partial_x u.$$

Therefore we can rewrite (2.2) in the following form

$$\begin{cases} \partial_v u = h(t, x, u), & t > 0, \ x \in \mathbb{R}, \\ u(0, x) = u_0(x), & x \in \mathbb{R}. \end{cases} \tag{2.3}$$

The idea consists in viewing (2.3) as an infinite family of Cauchy problems for ordinary differential equations parameterized on x.

Let $(\tau, \xi) \in [0, \infty) \times \mathbb{R}$ and consider the line

$$\gamma : [0, \infty) \longrightarrow [0, \infty) \times \mathbb{R}, \qquad \gamma(t) = (t - \tau)v + (\tau, \xi) = (t, (t - \tau)a + \xi).$$

Let u be a solution of (2.2) and consider its restriction to γ

$$v(t) = u(\gamma(t)) = u(t, (t - \tau)a + \xi).$$

We have that

$$\begin{aligned} v(\tau) =\ & u(\tau, \xi), \\ v'(t) =\ & (\partial_t u(\gamma(t)), \partial_x u(\gamma(t))) \cdot \gamma'(t) \\ =\ & \partial_v u(\gamma(t)) = h(\gamma(t), u(\gamma(t))) \\ =\ & h(\gamma(t), v(t)), \\ v(0) =\ & u(\gamma(0)) = u(0, \xi - a\tau) = u_0(\xi - a\tau). \end{aligned}$$

Since $v(\tau) = u(\tau, \xi)$, if we want to know the value of the solution to (2.2) in the point (τ, ξ) we have to solve the Cauchy problem

$$\begin{cases} v'(t) = h(t, (t - \tau)a + \xi, v(t)), \\ v(0) = u_0(\xi - a\tau), \end{cases}$$

and then evaluate v in τ.

Example 2.1 Consider the Cauchy problem

$$\begin{cases} \partial_t u + 2\partial_x u = 3u, & t > 0, \ x \in \mathbb{R}, \\ u(0, x) = \sin(x), & x \in \mathbb{R}. \end{cases} \tag{2.4}$$

Let $(\tau, \xi) \in [0, \infty) \times \mathbb{R}$ be fixed. The characteristic line through (τ, ξ) is

$$t \longmapsto (t, 2(t - \tau) + \xi).$$

Since the solution of

$$\begin{cases} v' = 3v, & t > 0, \\ v(0) = \sin(\xi - 2\tau), \end{cases}$$

is

$$v(t) = e^{3t} \sin(\xi - 2\tau)$$

and

$$v(\tau) = e^{3\tau} \sin(\xi - 2\tau),$$

the solution u of (2.4) is

$$u(t, x) = e^{3t} \sin(x - 2t), \qquad t \geq 0, \ x \in \mathbb{R}.$$

2.2 Hyperbolic Equations with Variable Coefficients

We are now ready for the formulation of the method of the characteristics for equations with coefficients depending on t and x.

Consider the Cauchy problem

$$\begin{cases} \partial_t u + a(t, x)\partial_x u = h(t, x, u), & t > 0,\ x \in \mathbb{R}, \\ u(0, x) = u_0(x), & x \in \mathbb{R}, \end{cases} \tag{2.5}$$

where $u_0 \in C^1(\mathbb{R})$, $a \in C^1([0, \infty) \times \mathbb{R})$, $h \in C^1([0, \infty) \times \mathbb{R}^2)$.

Let $(\tau, \xi) \in [0, \infty) \times \mathbb{R}$ and $x = x(t; \tau, \xi)$ be the solution of the Cauchy problem

$$\begin{cases} x' = a(t, x), \\ x(\tau) = \xi. \end{cases} \tag{2.6}$$

Let u be a solution of (2.5) and consider its restriction on the curve $t \mapsto (t, x(t; \tau, \xi))$

$$v(t) = u(t, x(t; \tau, \xi)).$$

We have that

$$v(\tau) = u(\tau, \xi),$$
$$v'(t) = \partial_t u(t, x(t; \tau, \xi)) + \partial_x u(t, x(t; \tau, \xi))x'(t; \tau, \xi)$$
$$= \partial_t u(t, x(t; \tau, \xi)) + \partial_x u(t, x(t; \tau, \xi))a(t, x(t; \tau, \xi))$$
$$= h(t, x(t; \tau, \xi), u(t, x(t; \tau, \xi)))$$
$$= h(t, x(t; \tau, \xi), v(t)),$$
$$v(0) = u(0, x(0; \tau, \xi)) = u_0(x(0; \tau, \xi)).$$

Since $v(\tau) = u(\tau, \xi)$, if we want to know the value of the solution to (2.5) in the point (τ, ξ) we have to solve the characteristic's Eq. (2.6), then the Cauchy problem

$$\begin{cases} v'(t) = h(t, x(t; \tau, \xi), v(t)), \\ v(0) = u_0(x(0; \tau, \xi)), \end{cases}$$

and finally evaluate v in τ.

Example 2.2 Consider the Cauchy problem

$$\begin{cases} \partial_t u + x\partial_x u = 2xu, & t > 0,\ x \in \mathbb{R}, \\ u(0, x) = 1, & x \in \mathbb{R}. \end{cases} \tag{2.7}$$

Let $(\tau, \xi) \in [0, \infty) \times \mathbb{R}$ be fixed. The characteristic's equation is

$$\begin{cases} x' = x, \\ x(\tau) = \xi, \end{cases}$$

therefore

$$x(t; \tau, \xi) = \xi e^{t-\tau}, \qquad t \geq 0.$$

Since the solution of

$$\begin{cases} v' = 2\xi e^{t-\tau} v, \quad t > 0, \\ v(0) = 1, \end{cases}$$

is

$$v(t) = e^{-2\xi e^{-\tau}} e^{2\xi e^{-\tau} e^{t}}$$

and

$$v(\tau) = e^{2\xi(1-e^{-\tau})},$$

the solution u of (2.7) is

$$u(t, x) = e^{2x(1-e^{-t})}, \qquad t \geq 0, \ x \in \mathbb{R}.$$

2.3 Hyperbolic Equations with Nonlinear Coefficients

We conclude this chapter showing that if the coefficients a depends on the unknown u the method of characteristics works only for short time due to the crossing of the characteristic lines.

Consider the Cauchy problem

$$\begin{cases} \partial_t u + a(t, x, u)\partial_x u = h(t, x, u), \quad t > 0, \ x \in \mathbb{R}, \\ u(0, x) = u_0(x), \qquad\qquad\qquad\quad x \in \mathbb{R}, \end{cases} \tag{2.8}$$

where $u_0 \in C^1(\mathbb{R})$, $a \in C^1([0, \infty) \times \mathbb{R}^2)$, $h \in C^1([0, \infty) \times \mathbb{R}^2)$ are Lipschitz continuous. For every $y \in \mathbb{R}$, (2.8) reduces to the system of ordinary differential equations

$$\begin{cases} v' = h(t, x, v), \\ x' = a(t, x, v), \\ v(0) = u_0(y), \\ x(0) = y, \end{cases} \tag{2.9}$$

whose solution is

$$t \longmapsto (x(t; y), v(t; y)).$$

Clearly

$$x(0; y) = y, \qquad v(0; y) = u_0(y).$$

The function

$$\sigma : (t, y) \in [0, \infty) \times \mathbb{R} \longmapsto (t, x(t; y), v(t; y)) \in \mathbb{R}^3$$

determines a bidimensional surface S of $\mathbb{R}^3$.

We claim that

$$\text{the surface } S \text{ is locally the graph} \tag{2.10}$$
$$\text{of a function } u = u(t, x) \text{ that solves (2.8).}$$

Due to the regularity of u_0, a, h we have $\sigma \in C^1([0, \infty) \times \mathbb{R}; \mathbb{R}^3)$. Let $x_0 \in \mathbb{R}$. The function

$$X : (t, y) \longmapsto (t, x(t; y))$$

is locally invertible in $(0, x_0)$. Indeed

$$\nabla_{(t,y)} X(t, y) = \begin{pmatrix} 1 & a(t, x(t; y), v(t; y)) \\ 0 & \partial_y x(t; y) \end{pmatrix},$$

$$\nabla_{(t,y)} X(0, x_0) = \begin{pmatrix} 1 & a(t, x_0, u_0(x_0)) \\ 0 & 1 \end{pmatrix},$$

$$\det \left(\nabla_{(t,y)} X(0, x_0) \right) = 1 (\neq 0).$$

Since X is invertible in a neighborhood Ω of $(0, x_0)$, S is locally the graph of a function $u = u(t, x)$ that belongs to $C^1(\Omega)$. We have to show that u solves (2.8).

Since

$$\sigma(0, x_0) = (0, x(0; x_0), v(0; x_0)) = (0, x_0, u_0(x_0)),$$

we have

$$u(0, x_0) = u_0(x_0). \tag{2.11}$$

Let $(\bar{t}, \bar{x}) \in \Omega$ and $\bar{y} \in \mathbb{R}$ such that

$$x(\bar{t}; \bar{y}) = \bar{x}.$$

Since

$$\sigma(\bar{t}, \bar{y}) = (\bar{t}, x(\bar{t}; \bar{y}), v(\bar{t}; \bar{y})) = (\bar{t}, \bar{x}, v(\bar{t}; \bar{y})),$$

we have

$$u(\bar{t}, \bar{x}) = v(\bar{t}; \bar{y})$$

and then

$$\begin{aligned}
&\partial_t u(\bar{t}, \bar{x}) + a(\bar{t}, \bar{x}, u(\bar{t}, \bar{x})) \partial_x u(\bar{t}, \bar{x}) \\
&= \partial_t u(\bar{t}, x(\bar{t}; \bar{y})) + a(\bar{t}, x(\bar{t}; \bar{y}), v(\bar{t}, \bar{y})) \partial_x u(\bar{t}, x(\bar{t}; \bar{y})) \\
&= \partial_t u(\bar{t}, x(\bar{t}; \bar{y})) + \left(\frac{d}{dt} x(\bar{t}; \bar{y}) \right) \partial_x u(\bar{t}, x(\bar{t}; \bar{y})) \\
&= \frac{d}{dt} u(\bar{t}, x(\bar{t}; \bar{y})) = \frac{d}{dt} v(\bar{t}; \bar{y}) \\
&= h(\bar{t}, x(\bar{t}; \bar{y}), v(\bar{t}; \bar{y})) = h(\bar{t}, \bar{x}, u(\bar{t}, \bar{x})),
\end{aligned}$$

that concludes the proof of (2.10).

We compare the properties of variable coefficients and nonlinear coefficient equations.

(1) In the variable coefficients case (see (2.6)) the characteristic curves are independent on the solution u and the initial condition u_0. Moreover, if the initial condition is supported in the interval $[a, b]$ then the solution u is supported in the set

$$\bigcup_{t \geq 0} \{t\} \times [x(t; 0, a), x(t; 0, b)],$$

that is independent on u and u_0. On the other hand, we can define the characteristics also in the nonlinear coefficient case (see (2.9)) but they depend on the solution u and the initial datum u_0.

(2) In both cases, if $x = x(t)$ is a characteristic curve, we have

$$\frac{d}{dt} u(t, x(t)) = h(t, x(t), u(t, x(t))).$$

Therefore, if u_0 is bounded and $h(t, x, \cdot)$ is sublinear then u stays bounded. On the contrary if $h(t, x, \cdot)$ is superlinear u may blow-up in finite time.

(3) In the variable coefficient case, as soon as u stays bounded then also $\partial_x u$ stays bounded. Indeed, $\partial_x u$ satisfies the equation

$$\partial_{tx}^2 u + a\partial_{xx}^2 u = (\partial_u h - \partial_x a)\partial_x u + \partial_x h,$$

that is linear in $\partial_x u$. Along a characteristic $x = x(t)$ we have

$$\frac{d}{dt} \partial_x u(t, x(t)) = (\partial_u h - \partial_x a)\partial_x u + \partial_x h,$$

that is a linear ordinary differential equation. Therefore, if u is bounded the same applies to $\partial_x u$. In the nonlinear coefficient case, $\partial_x u$ can blow-up in finite time. Indeed, $\partial_x u$ satisfies the equation

$$\partial_{tx}^2 u + a\partial_{xx}^2 u = -\partial_u a(\partial_x u)^2 + (\partial_u h - \partial_x a)\partial_x u + \partial_x h,$$

that is quadratic in $\partial_x u$. Along a characteristic $x = x(t)$ we have

$$\frac{d}{dt} \partial_x u(t, x(t)) = -\partial_u a(\partial_x u)^2 + (\partial_u h - \partial_x a)\partial_x u + \partial_x h,$$

that is a quadratic ordinary differential equation. Therefore, $|\partial_x u| \to \infty$ in finite time along a characteristic even if u stays bounded.

Example 2.3 Let us consider the Cauchy problem for the Burgers equation

$$\begin{cases} \partial_t u + u\partial_x u = 0, & t > 0, \ x \in \mathbb{R}, \\ u(0, x) = \dfrac{1}{1 + x^2}, & x \in \mathbb{R}. \end{cases} \tag{2.12}$$

For every $y \in \mathbb{R}$, (2.12) reduces to the system of ordinary differential equation

$$\begin{cases} v' = 0, \\ x' = v, \\ v(0) = \dfrac{1}{1+y^2}, \\ x(0) = y. \end{cases} \qquad (2.13)$$

The surface $\mathcal{S}$ is parametrized by the function

$$(t, y) \longmapsto \left(t, y + \frac{t}{1+y^2}, \frac{1}{1+y^2} \right).$$

For small t, the function

$$y \longmapsto x(t; y) = y + \frac{t}{1+y^2}$$

is invertible. Let $y = y(t; x)$ be its inverse. The solution of (2.12) is

$$u(t, x) = v(t; y(t; x)) = \frac{1}{1 + y^2(t; x)}.$$

We claim that $\partial_x u$ blows-up at a time $\bar{t} \leq \dfrac{8}{\sqrt{27}}$, therefore there are no smooth solutions of (2.12) for $t > \bar{t}$, see Fig. 2.1.

The equation satisfied by $\partial_x u$ is

$$\begin{cases} \partial_{tx}^2 u + u \partial_{xx}^2 u = -(\partial_x u)^2, & t > 0,\ x \in \mathbb{R}, \\ \partial_x u(0, x) = -\dfrac{2x}{(1+x^2)^2}, & x \in \mathbb{R}. \end{cases} \qquad (2.14)$$

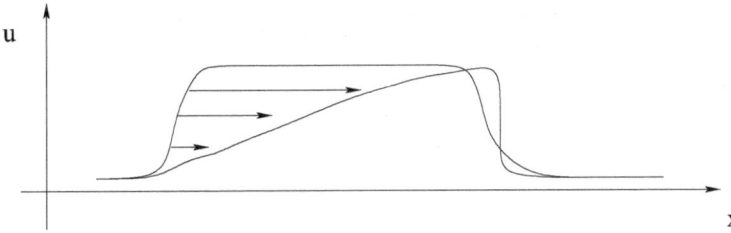

u

x

Fig. 2.1 Spontaneous creation of a shock

Along a characteristic $x = x(t; y)$ we have

$$\frac{d}{dt} \partial_x u(t, x(t; y)) = -(\partial_x u(t, x(t; y)))^2,$$

therefore

$$\partial_x u(t, x(t; y)) = \frac{1}{t + \dfrac{1}{\partial_x u(0, y)}}.$$

We consider one of the inflection points of u_0

$$y = \sqrt{\frac{1}{3}}.$$

We have

$$x\left(t; \sqrt{\frac{1}{3}}\right) = \sqrt{\frac{1}{3}} + \frac{t}{1 + \dfrac{1}{3}} = \sqrt{\frac{1}{3}} + \frac{3}{4}t,$$

$$\partial_x u\left(0, \sqrt{\frac{1}{3}}\right) = -\frac{2\sqrt{\dfrac{1}{3}}}{\left(1 + \dfrac{1}{3}\right)^2} = -\frac{\sqrt{27}}{8},$$

$$\partial_x u\left(t, x\left(t; \sqrt{\frac{1}{3}}\right)\right) = \frac{1}{t + \dfrac{1}{\partial_x u\left(0, \sqrt{\dfrac{1}{3}}\right)}} = \frac{1}{t - \dfrac{8}{\sqrt{27}}},$$

therefore

$$\lim_{t \to \left(\frac{8}{\sqrt{27}}\right)^-} \partial_x u\left(t, x\left(t; \sqrt{\frac{1}{3}}\right)\right) = \infty.$$

2.4 Exercises

Exercise 2.1 Using the method of characteristics, solve the following system of linear hyperbolic equations

$$\begin{cases} \partial_t u + \partial_x u = v, & t > 0,\ x \in \mathbb{R}, \\ \partial_t v - \partial_x v = 2u, & t > 0,\ x \in \mathbb{R}, \\ u(0, x) = v(0, x) = 1, & x \in \mathbb{R}. \end{cases} \quad (2.15)$$

Exercise 2.2 Using the method of characteristics, solve the following Cauchy problem

$$\begin{cases} \partial_t u + 5\partial_x u = u, & t > 0,\ x \in \mathbb{R}, \\ u(0, x) = e^x, & x \in \mathbb{R}. \end{cases} \quad (2.16)$$

Exercise 2.3 Using the method of characteristics, solve the following Cauchy problem

$$\begin{cases} \partial_t u + x\partial_x u = x, & t > 0,\ x \in \mathbb{R}, \\ u(0, x) = \arctan(x), & x \in \mathbb{R}. \end{cases} \quad (2.17)$$

Exercise 2.4 Using the method of characteristics, prove that the solution of the following Cauchy problem does not develop discontinuities in finite time

$$\begin{cases} \partial_t u + u\partial_x u = 0, & t > 0,\ x \in \mathbb{R}, \\ u(0, x) = \arctan(x), & x \in \mathbb{R}. \end{cases} \quad (2.18)$$

Reference

1. Evans, L.C.: Partial Differential Equations. Graduate Studies in Mathematics, vol. 19, 2nd edn. American Mathematical Society, Providence (2010)

Chapter 3
Entropy Solutions

Abstract This chapter is devoted to the concept of weak entropy solution. We give the definition of distributional solution and prove that a shock is a distributional solution if and only if the Rankine-Hugoniot Condition holds. We then provide the definition of entropy solution and prove a necessary and sufficient condition for a shock to be an entropy solution. We conclude the chapter with the Kružkov Theorem stating the uniqueness and stability the entropy solution of a Cauchy problem.

Keywords Cauchy problem · Weak solutions · Rankine-Hugoniot condition · Nonuniqueness of weak solutions · Entropies · Entropy fluxes · Kružkov entropies · Entropy solutions · Kružkov Theorem · Doubling of variables · Uniqueness · Stability

We showed in Chap. 2 that even a Cauchy problem of the type

$$\partial_t u + \partial_x \left(\frac{u^2}{2} \right) = 0, \qquad u(0, x) = \frac{1}{1 + x^2},$$

with analytic flux ($u \mapsto u^2/2$) and analytic initial condition ($x \mapsto 1/(1 + x^2)$) may experience discontinuities in finite time. As a consequence we develop a wellposedness theory for conservation laws in the framework of *entropy solutions*, which are special distributional solutions satisfying additional inequalities. The definition is inspired by the Second Principle of Thermodynamics, we consider only the distributional solutions along which the entropies decrease. Note that the physical entropies are all concave maps, in the mathematical community the entropies are assumed to be convex, this explain the discrepancy between the usual Second Principle of Thermodynamics and the definition of entropy solution.

G. M. Coclite, *Scalar Conservation Laws*, SpringerBriefs in Mathematics,
https://doi.org/10.1007/978-981-97-3984-4_3

3.1 Weak Solutions

Consider the conservation law

$$\partial_t u + \partial_x f(u) = 0, \qquad t > 0, \ x \in \mathbb{R}, \tag{3.1}$$

endowed with the initial condition

$$u(0, x) = u_0(x), \qquad x \in \mathbb{R}, \tag{3.2}$$

and assume

$$f \in C^2(\mathbb{R}), \qquad u_0 \in L^\infty_{loc}(\mathbb{R}). \tag{3.3}$$

Definition 3.1 A function $u : [0, \infty) \times \mathbb{R} \to \mathbb{R}$ is a weak solution of the Cauchy problem (3.1) and (3.2), if

(i) $u \in L^\infty_{loc}((0, \infty) \times \mathbb{R})$;
(ii) u satisfies (3.1) and (3.2) in the sense of distributions in $[0, \infty) \times \mathbb{R}$, namely for every test function $\varphi \in C^\infty(\mathbb{R}^2)$ with compact support we have

$$\int_0^\infty \int_{\mathbb{R}} (u \partial_t \varphi + f(u) \partial_x \varphi) \, dt dx + \int_{\mathbb{R}} u_0(x) \varphi(0, x) dx = 0.$$

We say that u is a weak solution of the conservation law (3.1) if (i) holds and

(iii) u satisfies (3.1) in the sense of distributions in $(0, \infty) \times \mathbb{R}$, namely for every test function $\varphi \in C^\infty((0, \infty) \times \mathbb{R})$ with compact support we have

$$\int_0^\infty \int_{\mathbb{R}} (u \partial_t \varphi + f(u) \partial_x \varphi) \, dt dx = 0.$$

Direct consequence of the Dominated Convergence Theorem is the following.

Theorem 3.1 *Let* $\{u_\varepsilon\}_{\varepsilon > 0}$ *and* u *be functions defined on* $[0, \infty) \times \mathbb{R}$ *with values in* $\mathbb{R}$. *If*

(i) *there exists* $M > 0$ *such that* $\|u_\varepsilon\|_{L^\infty((0,\infty) \times \mathbb{R})} \leq M$ *for every* $\varepsilon > 0$;
(ii) $u \in L^\infty((0, \infty) \times \mathbb{R})$;
(iii) $u_\varepsilon \to u$ *in* $L^1_{loc}((0, \infty) \times \mathbb{R})$ *as* $\varepsilon \to 0$;
(iv) *every* u_ε *is a weak solution of* (3.1);

then u *is a weak solution of* (3.1).

3.2 Rankine-Hugoniot Condition

In this section we analyze the *shocks*, which are the simplest discontinuous weak solutions of (3.1).

Let u_-, u_+, $\lambda \in \mathbb{R}$ be fixed and consider the function

$$U : [0, \infty) \times \mathbb{R} \longrightarrow \mathbb{R}, \qquad U(t, x) = \begin{cases} u_-, & \text{if } x < \lambda t, \\ u_+, & \text{if } x \geq \lambda t. \end{cases} \tag{3.4}$$

Since we are not interested to the trivial case $u_+ = u_-$ in the following we always assume

$$u_+ \neq u_-.$$

Theorem 3.2 (Rankine-Hugoniot Condition) *The following statements are equivalent:*

(i) the function U defined in (3.4) is a weak solution of (3.1);
(ii) the Rankine-Hugoniot condition holds, i.e.,

$$f(u_+) - f(u_-) = \lambda(u_+ - u_-). \tag{3.5}$$

Remark 3.1 The Rankine-Hugoniot Condition (3.5) is a scalar equation that links the right and left sates u_+, u_- and the speed λ of the shock. In particular, if f is Lipschitz continuous with Lipschitz constant L, (3.5) gives

$$|\lambda| = \frac{|f(u_+) - f(u_-)|}{|u_+ - u_-|} \leq L.$$

In other terms, the speed of propagation of the singularities is finite and varies between $-L$ and L.

Proof of Theorem 3.2 Let $\varphi \in C^\infty((0, \infty) \times \mathbb{R})$ be a test function with compact support. Consider the vector field

$$F = (U\varphi, f(U)\varphi)$$

and the domains

$$\Omega_+ = \{x > \lambda t\}, \qquad \Omega_- = \{x < \lambda t\}.$$

The definition of U gives

$$(t, x) \in \Omega_+ \implies \begin{cases} F(t, x) = (u_+\varphi, f(u_+)\varphi), \\ \operatorname{div}_{(t,x)}(F)(t, x) = u_+\partial_t\varphi + f(u_+)\partial_x\varphi, \end{cases}$$

$$(t, x) \in \Omega_- \implies \begin{cases} F(t, x) = (u_-\varphi, f(u_-)\varphi), \\ \operatorname{div}_{(t,x)}(F)(t, x) = u_-\partial_t\varphi + f(u_-)\partial_x\varphi. \end{cases}$$

Since

$$\partial\Omega_+ = \partial\Omega_- = \{x = \lambda t\},$$

and the outer normals to Ω_+ and Ω_- are $(\lambda, -1)$ and $(-\lambda, 1)$ we have

$$\int_0^\infty \int_{\mathbb{R}} (U\partial_t\varphi + f(U)\partial_x\varphi)dtdx$$

$$= \iint_{\Omega_+} (u_+\partial_t\varphi + f(u_+)\partial_x\varphi)dtdx$$

$$+ \iint_{\Omega_-} (u_-\partial_t\varphi + f(u_-)\partial_x\varphi)dtdx$$

$$= \iint_{\Omega_+} \operatorname{div}(F)dtdx + \iint_{\Omega_-} \operatorname{div}(F)dtdx$$

$$= \int_0^\infty (u_+, f(u_+)) \cdot (\lambda, -1)\varphi(t, \lambda t)dt$$

$$+ \int_0^\infty (u_-, f(u_-)) \cdot (-\lambda, 1)\varphi(t, \lambda t)dt$$

$$= [\lambda(u_+ - u_-) - (f(u_+) - f(u_-))] \int_0^\infty \varphi(t, \lambda t)dt.$$

Therefore

$$\int_0^\infty \int_{\mathbb{R}} (U\partial_t\varphi + f(U)\partial_x\varphi)dtdx = 0, \quad \forall\varphi$$

$$\Updownarrow$$

$$f(u_+) - f(u_-) = \lambda(u_+ - u_-).$$

$\square$

Theorem 3.3 *Let* $u : [0, \infty) \times \mathbb{R} \to \mathbb{R}$, $\tau > 0$, $\xi \in \mathbb{R}$. *If*

(i) $u \in L^\infty_{loc}((0, \infty) \times \mathbb{R})$;

(ii) u is a weak solution of (3.1);

(iii) $\displaystyle \lim_{\varepsilon \to 0} \frac{1}{\varepsilon^2} \int_{-\varepsilon}^{\varepsilon} \int_{-\varepsilon}^{\varepsilon} |u(t + \tau, x + \xi) - U(t, x)| dt dx = 0;$

then (3.5) holds.

Proof For every $\mu > 0$ define

$$u_\mu(t, x) = u(\mu t + \tau, \mu x + \xi), \qquad t \geq -\frac{\tau}{\mu}, \ x \in \mathbb{R}.$$

Since u is a weak solution of (3.1), the same does u_μ. We claim that

$$u_\mu \longrightarrow U, \qquad f(u_\mu) \longrightarrow f(U), \qquad \text{in } L^1_{loc}((0, \infty) \times \mathbb{R}), \text{ as } \mu \to 0. \qquad (3.6)$$

Let $R > 0$ and $\mu < \frac{\tau}{R}$. Since

$$U(\mu t, \mu x) = U(t, x), \qquad t > 0, \ x \in \mathbb{R},$$

we get

$$\int_{-R}^{R} \int_{-R}^{R} |u_\mu(t, x) - U(t, x)| dt dx$$

$$= \frac{1}{\mu^2} \int_{-R\mu}^{R\mu} \int_{-R\mu}^{R\mu} |u(t + \tau, x + \xi) - U(t, x)| dt dx \longrightarrow 0,$$

namely

$$u_\mu \longrightarrow U, \qquad \text{in } L^1((-R, R) \times (-R, R)), \text{ as } \mu \to 0.$$

Therefore the Dominated Convergence Theorem gives (3.6). Finally, Theorem 3.1 and (3.6) implies that U is a weak solution of (3.1). Then, the claim follows from Theorem 3.2. $\qquad \qquad \qquad \qquad \qquad \qquad \qquad \qquad \qquad \qquad \qquad \qquad \quad \Box$

3.3 Nonuniqueness of Weak Solutions

In this section, we show with a simple example that the Cauchy problem (3.1)–(3.2) may admit more than one weak solution.

Let us consider the Riemann problem for the Bugers equation

$$\partial_t u + \partial_x \left(\frac{u^2}{2} \right) = 0, \qquad u(0, x) = \begin{cases} 0, & \text{if } x < 0, \\ 1, & \text{if } x \geq 0. \end{cases} \qquad (3.7)$$

Thanks to Theorem 3.2, we know that the function

$$U(t, x) = \begin{cases} 0, & \text{if } x < \frac{1}{2}t, \\ 1, & \text{if } x \geq \frac{1}{2}t, \end{cases}$$

is a weak solution of (3.7).

Consider the function

$$v(t, x) = \begin{cases} 0, & \text{if } x < 0, \\ \dfrac{x}{t}, & \text{if } 0 \leq x \leq t, \\ 1, & \text{if } x \geq t. \end{cases}$$

Since for every test function $\varphi \in C^\infty(\mathbb{R}^2)$ with compact support

$$\int_0^\infty \int_{\mathbb{R}} \left(v \partial_t \varphi + \frac{v^2}{2} \partial_x \varphi \right) dt\, dx + \int_0^\infty \varphi(0, x) dx$$

$$= \int_0^\infty \left(\int_x^\infty \frac{x}{t} \partial_t \varphi\, dt \right) dx + \int_0^\infty \left(\int_0^t \frac{x^2}{2t^2} \partial_x \varphi\, dx \right) dt$$

$$+ \int_0^\infty \left(\int_0^x \partial_t \varphi\, dt \right) dx + \int_0^\infty \left(\int_t^\infty \partial_x \varphi\, dx \right) dt + \int_0^\infty \varphi(0, x) dx = 0,$$

v is a also a weak solution of (3.7).

3.4 Entropy Conditions

We showed in the previous section that the Cauchy problem (3.1)–(3.2) may admit more than one weak solution. In this section we introduce some additional conditions that will select the unique "physically meaningful" solution within the family of the weak solutions. Those conditions are inspired by the Second Principle of Thermodynamics.

Definition 3.2 Let $\eta, q : \mathbb{R} \rightarrow \mathbb{R}$ be functions. We say that η is an entropy associated to (3.1) with flux q if

$$\eta, q \in C^2(\mathbb{R}), \qquad \eta'' \geq 0, \qquad \eta' f' = q'.$$

Remark 3.2 If u is a classical solution of (3.1) and η is an entropy with flux q we have

$$\partial_t \eta(u) + \partial_x q(u) = 0.$$

Indeed,

$$\partial_t \eta(u) + \partial_x q(u) = \eta'(u)\partial_t u + q'(u)\partial_x u$$
$$= \eta'(u)\left(\partial_t u + f'(u)\partial_x u\right)$$
$$= \eta'(u)\left(\partial_t u + \partial_x f(u)\right) = 0.$$

Unfortunately, if u is discontinuous the chain rule does not work and the previous argument cannot be used.

Definition 3.3 A function $u : [0, \infty) \times \mathbb{R} \to \mathbb{R}$ is an entropy solution of the Cauchy problem (3.1) and (3.2), if

(i) $u \in L^{\infty}_{loc}((0, \infty) \times \mathbb{R})$;
(ii) for every entropy η with flux q, u satisfies

$$\partial_t \eta(u) + \partial_x q(u) \le 0, \qquad \eta(u(0, \cdot)) = \eta(u_0), \qquad (3.8)$$

in the sense of distributions in $[0, \infty) \times \mathbb{R}$, namely for every nonnegative test function $\varphi \in C^{\infty}(\mathbb{R}^2)$ with compact support we have

$$\int_0^{\infty} \int_{\mathbb{R}} (\eta(u)\partial_t \varphi + q(u)\partial_x \varphi)\, dtdx + \int_{\mathbb{R}} \eta(u_0(x))\varphi(0, x)dx \ge 0. \qquad (3.9)$$

We say that u is an entropy solution of the conservation law (3.1) if (i) holds and

(iii) for every entropy η with flux q, u satisfies

$$\partial_t \eta(u) + \partial_x q(u) \le 0 \qquad (3.10)$$

in the sense of distributions in $(0, \infty) \times \mathbb{R}$, namely for every nonnegative test function $\varphi \in C^{\infty}((0, \infty) \times \mathbb{R})$ with compact support we have

$$\int_0^{\infty} \int_{\mathbb{R}} (\eta(u)\partial_t \varphi + q(u)\partial_x \varphi)\, dtdx \ge 0.$$

Even if these definitions seem in contradiction with the Second Principle of Thermodynamics they are not because the physical entropies are concave while the one we are using are convex.

As a direct consequence of the Dominated Convergence Theorem, we have the following.

Theorem 3.4 *Let $\{u_\varepsilon\}_{\varepsilon>0}$ and u be functions defined on $[0, \infty) \times \mathbb{R}$ with values in $\mathbb{R}$. If*

   (i) *there exists $M > 0$ such that $\|u_\varepsilon\|_{L^\infty((0,\infty)\times\mathbb{R})} \leq M$ for every $\varepsilon > 0$;*
  (ii) *$u \in L^\infty((0, \infty) \times \mathbb{R})$;*
 (iii) *$u_\varepsilon \to u$ in $L^1_{loc}((0, \infty) \times \mathbb{R})$ as $\varepsilon \to 0$;*
 (iv) *every u_ε is a entropy solution of (3.1);*

then

$$u \text{ is a entropy solution of (3.1)}.$$

A fundamental class of entropies are the ones introduced by Kružkov [1]

$$\eta(\xi) = |\xi - c|, \qquad q(\xi) = \text{sign}\,(\xi - c)\,(f(\xi) - f(c)), \qquad \xi \in \mathbb{R}, \qquad (3.11)$$

for every constant $c \in \mathbb{R}$.

Since the Kružkov entropies are not C^2, the following theorem is needed.

Theorem 3.5 *Let $u : [0, \infty) \times \mathbb{R} \to \mathbb{R}$ be a function. If*

$$u \in L^\infty_{loc}((0, \infty) \times \mathbb{R}),$$

then the following statements are equivalent

  (i) *u is an entropy solution of (3.1)–(3.2);*
 (ii) *for every $c \in \mathbb{R}$ and every nonnegative test function $\varphi \in C^\infty(\mathbb{R}^2)$ with compact support*

$$\int_0^\infty \int_\mathbb{R} (|u - c|\partial_t\varphi + \text{sign}\,(u - c)\,(f(u) - f(c))\partial_x\varphi)\,dtdx$$

$$+ \int_\mathbb{R} |u_0(x) - c|\varphi(0, x)dx \geq 0. \qquad (3.12)$$

Remark 3.3 The set of the entropies

$$\{\eta \in C^2(\mathbb{R}); \eta \text{ convex}\}$$

is an infinite dimensional manifold. On the other hand, the set of the Kružkov entropies

$$\{|\cdot -c|; c \in \mathbb{R}\}$$

is a one-dimensional manifold. Therefore the previous theorem says that if we have to verify that a function is an entropy solution of (3.1) we can use just the Kružkov entropies and the "amount" of inequalities to verify is "much less" than the one required in Definition 3.3.

Proof of Theorem 3.5 Let us start by proving (i) $\Rightarrow$ (ii). Let $c \in \mathbb{R}$ and $\varphi \in C^\infty(\mathbb{R}^2)$ be a nonnegative test function with compact support. For every $n \in \mathbb{N} \setminus \{0\}$, consider the functions

$$\eta_n(\xi) = \sqrt{(\xi - c)^2 + \frac{1}{n}}, \qquad q_n(\xi) = \int_c^\xi \frac{\sigma - c}{\sqrt{(\sigma - c)^2 + \frac{1}{n}}} f'(\sigma) d\sigma, \qquad \xi \in \mathbb{R}.$$

Since

$$\eta_n \in C^2(\mathbb{R}),$$

$$\eta_n'(\xi) = \frac{\xi - c}{\sqrt{(\xi - c)^2 + \frac{1}{n}}},$$

$$\eta_n''(\xi) = \frac{1}{n\left((\xi - c)^2 + \frac{1}{n}\right)^{\frac{3}{2}}} \geq 0,$$

$$q_n' = \eta_n' f',$$

we have

$$\int_0^\infty \int_{\mathbb{R}} \left(\eta_n(u)\partial_t \varphi + q_n(u)\partial_x \varphi\right) dt dx + \int_{\mathbb{R}} \eta_n(u_0(x))\varphi(0, x) dx \geq 0.$$

As $n \to \infty$ thanks to the Dominated Convergence Theorem we get (3.12).

Let us prove (ii) $\Rightarrow$ (i). Let η be an entropy with flux q and $\varphi \in C^\infty(\mathbb{R}^2)$ be a nonnegative test function with compact support. Define

$$M = \sup_{\text{supp}(\varphi)} |u|.$$

We approximate η' with piecewise constant functions in $[-M, M]$. For every $n \in \mathbb{N} \setminus \{0\}$, consider

$$\eta_n(\xi) = \int_{-M}^\xi k_n(\sigma) d\sigma + \eta(-M),$$

$$k_n(\sigma) = \sum_{j=0}^{2n-1} \eta'\left(\frac{M}{n}j - M\right) \chi_{\left[\frac{M}{n}j - M, \frac{M}{n}(j+1) - M\right]}(\sigma),$$

$$q_n(\xi) = \int_{-M}^\xi f'(\sigma)k_n(\sigma) d\sigma.$$

We have

$$k_n(\sigma) = \sum_{j=0}^{n-1} a_j \left[\operatorname{sign}(\sigma - b_j) + c_j\right] \chi_{\left[2\frac{M}{n}j - M, 2\frac{M}{n}(j+1) - M\right)}(\sigma),$$

where

$$a_j = \frac{1}{2}\left(\eta'\left(\frac{M}{n}(2j+1) - M\right) - \eta'\left(\frac{M}{n}2j - M\right)\right),$$

$$b_j = \frac{M}{n}(2j+1) - M,$$

$$c_j = \frac{1}{2}\left(\eta'\left(\frac{M}{n}(2j+1) - M\right) + \eta'\left(\frac{M}{n}2j - M\right)\right).$$

Since $\eta'' \geq 0$ we have $a_j \geq 0$ and then

$$\int_0^\infty \int_{\mathbb{R}} (\eta_n(u)\partial_t\varphi + q_n(u)\partial_x\varphi)\,dt\,dx + \int_{\mathbb{R}} \eta_n(u_0(x))\varphi(0, x)\,dx \geq 0.$$

As $n \to \infty$ thanks to the Dominated Convergence Theorem we get (3.9).   □

It is clear that a smooth solutions is both an entropy and a weak solution (see Remark 3.2). We conclude this section proving that the entropy solutions are weak solutions. In the next section we will show that there are weak solution that are not entropy ones.

Theorem 3.6 *Let $u : [0, \infty) \times \mathbb{R} \to \mathbb{R}$ be a function. If*

$$u \in L^\infty_{loc}((0, \infty) \times \mathbb{R}),$$

and u is an entropy solution of (3.1)–(3.2) then u is a weak solution of (3.1)–(3.2).

Proof Let $\varphi \in C^2(\mathbb{R}^2)$ be a test function with compact support. Define

$$\varphi_+ = \max\{\varphi, 0\}, \qquad \varphi_- = \max\{-\varphi, 0\},$$

clearly

$$\varphi = \varphi_+ - \varphi_-, \qquad \varphi_+, \varphi_- \geq 0.$$

Using a smooth approximation of $\varphi_\pm$ and then passing to the limit we get

$$\int_0^\infty \int_{\mathbb{R}} (|u - c|\partial_t\varphi_\pm + \text{sign}\,(u - c)\,(f(u) - f(c))\partial_x\varphi_\pm)\,dt\,dx$$

$$+ \int_{\mathbb{R}} |u_0(x) - c|\varphi_\pm(0, x)dx \geq 0, \tag{3.13}$$

for every $c \in \mathbb{R}$.
   Define

$$M = \sup_{\text{supp}(\varphi)} |u|.$$

Choosing $c = M + 1$ in (3.13) we get

$$\int_0^\infty \int_{\mathbb{R}} ((M + 1 - u)\partial_t\varphi_\pm + (f(M + 1) - f(u))\partial_x\varphi_\pm)\,dt\,dx$$

$$+ \int_{\mathbb{R}} (M + 1 - u_0(x))\varphi_\pm(0, x)dx \geq 0,$$

and integrating by parts (since $M + 1$ is a classical solution of (3.1)) we get

$$\int_0^\infty \int_{\mathbb{R}} (u\partial_t\varphi_\pm + f(u)\partial_x\varphi_\pm)\,dt\,dx + \int_{\mathbb{R}} u_0(x)\varphi_\pm(0, x)dx \leq 0. \tag{3.14}$$

On the other hand, if we choose $c = -M - 1$ in (3.13) we get

$$\int_0^\infty \int_{\mathbb{R}} ((u + M + 1)\partial_t\varphi_\pm + (f(u) - f(-M - 1))\partial_x\varphi_\pm)\,dt\,dx$$

$$+ \int_{\mathbb{R}} (u_0(x) + M + 1)\varphi_\pm(0, x)dx \geq 0,$$

and integrating by parts (since $-M - 1$ is a classical solution of (3.1)) we get

$$\int_0^\infty \int_{\mathbb{R}} (u\partial_t\varphi_\pm + f(u)\partial_x\varphi_\pm)\,dt\,dx + \int_{\mathbb{R}} u_0(x)\varphi_\pm(0, x)dx \geq 0. \tag{3.15}$$

Adding (3.14) and (3.15) we get (3.9).   □

3.5 Entropic Shocks

In Sect. 3.2 we introduced the shock U (see (3.4)) and proved that it is a weak solution of (3.1) if and only if the Rankine-Hugoniot Condition (3.5) holds. In this section we prove a similar result giving a necessary and sufficient condition for the shock to be an entropy solution.

Theorem 3.7 (Liu-Ruggeri [2]) *The following statements are equivalent:*

(i) the function U defined in (3.4) is an entropy solution of (3.1);
(ii) the Rankine-Hugoniot Condition holds, i.e.,

$$f(u_+) - f(u_-) = \lambda(u_+ - u_-), \tag{3.16}$$

and

$$\begin{cases} f(\theta u_+ + (1-\theta)u_-) \geq \theta f(u_+) + (1-\theta)f(u_-), & \text{if } u_- < u_+, \\ f(\theta u_+ + (1-\theta)u_-) \leq \theta f(u_+) + (1-\theta)f(u_-), & \text{if } u_- > u_+, \end{cases} \tag{3.17}$$

for every $0 < \theta < 1$.

The inequalities in (3.17) have a simple geometric interpretation. If $u_- < u_+$ the graph of f has to be above the segment connecting $(u_-, f(u_-))$ and $(u_+, f(u_+))$, that is always true if f is concave. On the other hand if $u_- > u_+$ the graph of f has to be below the segment connecting $(u_+, f(u_+))$ and $(u_-, f(u_-))$, that is always true if f is convex. In particular, if f is concave the entropic shocks are upward and if is convex they are downward.

Moreover, we can rewrite (3.17) in the following way

$$\frac{f(u_*) - f(u_-)}{u_* - u_-} \geq \frac{f(u_+) - f(u_*)}{u_+ - u_*}, \tag{3.18}$$

for every $\min\{u_+, u_-\} < u_* < \max\{u_+, u_-\}$.

Indeed, if $u_- < u_+$ (in the case $u_- > u_+$ the same argument works) and $u_* = \theta u_+ + (1-\theta)u_-$ for some $0 < \theta < 1$ we have

$$\frac{f(u_*) - f(u_-)}{u_* - u_-} - \frac{f(u_+) - f(u_*)}{u_+ - u_*}$$

$$= \frac{f(u_*)(u_+ - u_-)}{(u_* - u_-)(u_+ - u_*)} - \frac{f(u_-)}{u_* - u_-} - \frac{f(u_+)}{u_+ - u_*}$$

$$\geq \frac{(\theta f(u_+) + (1-\theta)f(u_-))(u_+ - u_-)}{(u_* - u_-)(u_+ - u_*)} - \frac{f(u_-)}{u_* - u_-} - \frac{f(u_+)}{u_+ - u_*}$$

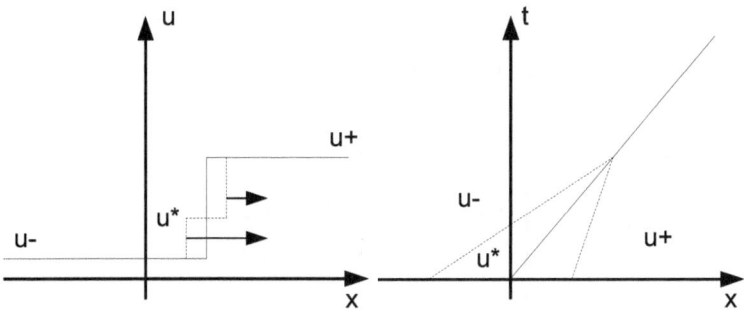

Fig. 3.1 Perturbation of an entropic shock

$$= f(u_+)\frac{\theta(u_+ - u_-) - (u_* - u_-)}{(u_* - u_-)(u_+ - u_*)}$$

$$+ f(u_-)\frac{(1 - \theta)(u_+ - u_-) - (u_+ - u_*)}{(u_* - u_-)(u_+ - u_*)} = 0.$$

We can interpret (3.18) as a stability condition. Indeed, if $u_- < u_* < u_+$ we can perturb the shock (u_-, u_+) and split it in the two shocks (u_-, u_*), (u_*, u_+). The two quantities in (3.18) give the speed of these two shocks: the one on the left is faster than the one on the right. Then the two waves will interact in finite time and generate again the initial shock (u_-, u_+) (see Fig. 3.1).

Lemma 3.1 *The following statements are equivalent:*

(i) the function U defined in (3.4) is an entropy solution of (3.1);
(ii) for every entropy η with flux q the following inequlity holds

$$\lambda(\eta(u_+) - \eta(u_-)) \geq q(u_+) - q(u_-); \tag{3.19}$$

(iii) for every constant $c \in \mathbb{R}$

$$\lambda(|u_+ - c| - |u_- - c|)$$
$$\geq \text{sign}\,(u_+ - c)\,(f(u_+) - f(c)) \tag{3.20}$$
$$- \text{sign}\,(u_- - c)\,(f(u_-) - f(c)).$$

Proof Let $\varphi \in C^\infty((0, \infty) \times \mathbb{R})$ be a nonnegative test function with compact support and η be an entropy with flux q. Consider the vector field

$$G = (\eta(U)\varphi, q(U)\varphi),$$

and the domains

$$\Omega_+ = \{x > \lambda t\}, \qquad \Omega_- = \{x < \lambda t\}.$$

The definition of U gives

$$(t, x) \in \Omega_+ \implies \begin{cases} G(t, x) = (\eta(u_+)\varphi, q(u_+)\varphi), \\ \text{div}_{(t,x)}(G)(t, x) = \eta(u_+)\partial_t\varphi + q(u_+)\partial_x\varphi, \end{cases}$$

$$(t, x) \in \Omega_- \implies \begin{cases} G(t, x) = (\eta(u_-)\varphi, q(u_-)\varphi), \\ \text{div}_{(t,x)}(G)(t, x) = \eta(u_-)\partial_t\varphi + q(u_-)\partial_x\varphi. \end{cases}$$

Since

$$\partial\Omega_+ = \partial\Omega_- = \{x = \lambda t\},$$

and the outer normals to Ω_+ and Ω_- are $(\lambda, -1)$ and $(-\lambda, 1)$ we have

$$\int_0^\infty \int_{\mathbb{R}} (\eta(U)\partial_t\varphi + q(U)\partial_x\varphi)dt dx$$

$$= \iint_{\Omega_+} (\eta(u_+)\partial_t\varphi + q(u_+)\partial_x\varphi)dt dx$$

$$+ \iint_{\Omega_-} (\eta(u_-)\partial_t\varphi + q(u_-)\partial_x\varphi)dt dx$$

$$= \iint_{\Omega_+} \text{div}(G)dt dx + \iint_{\Omega_-} \text{div}(G)dt dx$$

$$= \int_0^\infty (\eta(u_+), q(u_+)) \cdot (\lambda, -1)\varphi(t, \lambda t)dt$$

$$+ \int_0^\infty (\eta(u_-), q(u_-)) \cdot (-\lambda, 1)\varphi(t, \lambda t)dt$$

$$= [\lambda(\eta(u_+) - \eta(u_-)) - (q(u_+) - q(u_-))] \int_0^\infty \varphi(t, \lambda t)dt.$$

Therefore

$$\int_0^\infty \int_{\mathbb{R}} (\eta(U)\partial_t\varphi + q(U)\partial_x\varphi)dt dx \geq 0, \quad \forall\, \varphi$$

$$\Updownarrow$$

$$\lambda(\eta(u_+) - \eta(u_-)) \geq q(u_+) - q(u_-).$$

Therefore we proved that (i) $\Leftrightarrow$ (ii). The same argument works for (i) $\Leftrightarrow$ (iii). $\square$

Proof of Theorem 3.7 We begin by proving that (i) $\Rightarrow$ (ii). Since U is an entropy solution of (3.1), Theorem 3.2 gives (3.16). We have to prove (3.17). We distinguish two cases. We assume $u_- < u_+$. Let $0 < \theta < 1$ be fixed. We choose

$$c = \theta u_+ + (1 - \theta)u_-.$$

Since

$$u_- < c < u_+,$$

(3.20) gives

$$f(u_+) + f(u_-) - 2f(c) \leq \lambda(u_+ + u_- - 2c). \tag{3.21}$$

Using (3.16) and (3.21)

$$
\begin{aligned}
2f(\theta u_+ + (1 - \theta)u_-) &= 2f(c) \\
&\geq f(u_+) + f(u_-) - \lambda(u_+ + u_- - 2c) \\
&= f(u_+) + f(u_-) - \lambda(u_+ + u_- - 2(\theta u_+ + (1 - \theta)u_-)) \\
&= f(u_+) + f(u_-) - \lambda(1 - 2\theta)(u_+ - u_-) \\
&= f(u_+) + f(u_-) - (1 - 2\theta)(f(u_+) - f(u_-)) \\
&= 2(\theta f(u_+) + (1 - \theta)f(u_-)).
\end{aligned}
$$

Since the case $u_+ < u_+$ is analogous (3.17) is proved.

We have to prove that (ii) $\Rightarrow$ (i). It is enough to verify that (3.20) holds for every $c \in \mathbb{R}$. We distinguish four cases.

If

$$c \leq \min\{u_+, u_-\},$$

(3.16) gives

$$
\begin{aligned}
\lambda(|u_+ - c| - |u_- - c|) &= \lambda(u_+ - u_-) \\
&= f(u_+) - f(u_-) = (f(u_+) - f(c)) - (f(u_-) - f(c)) \\
&= \operatorname{sign}(u_+ - c)(f(u_+) - f(c)) - \operatorname{sign}(u_- - c)(f(u_-) - f(c)).
\end{aligned}
$$

If

$$c \geq \max\{u_+, u_-\},$$

the same argument applies.

If

$$u_- < c < u_+,$$

there exists $0 < \theta < 1$ such that

$$c = \theta u_+ + (1 - \theta)u_-.$$

(3.17) guarantees

$$f(c) \geq \theta f(u_+) + (1 - \theta)f(u_-),$$

then using (3.16)

$$
\begin{aligned}
\lambda(|u_+ - c| - |u_- - c|) &= \lambda(u_+ + -u_- - 2c) \\
&= \lambda(1 - 2\theta)(u_+ - u_-) = (1 - 2\theta)(f(u_+) - f(u_-)) \\
&= f(u_+) + f(u_-) - 2(\theta f(u_+) + (1 - \theta)f(u_-)) \\
&\geq f(u_+) + f(u_-) - 2f(c) \\
&= \operatorname{sign}(u_+ - c)(f(u_+) - f(c)) - \operatorname{sign}(u_- - c)(f(u_-) - f(c)).
\end{aligned}
$$

Finally, if

$$u_- < c < u_+,$$

the same argument works. Then (3.20) holds for every $c \in \mathbb{R}$.   □

Theorem 3.8 *Let* $u : [0, \infty) \times \mathbb{R} \to \mathbb{R}$, $\tau > 0$, $\xi \in \mathbb{R}$. *If*

(a) $u \in L^\infty_{loc}((0, \infty) \times \mathbb{R})$;
(b) u *is an entropy solution of* (3.1);
(c) $\displaystyle \lim_{\varepsilon \to 0} \frac{1}{\varepsilon^2} \int_{-\varepsilon}^{\varepsilon} \int_{-\varepsilon}^{\varepsilon} |u(t + \tau, x + \xi) - U(t, x)| dt dx = 0$;

then (3.16) *and* (3.17) *hold*.

Proof For every $\mu > 0$ define

$$u_\mu(t, x) = u(\mu t + \tau, \mu x + \xi), \qquad t \geq -\frac{\tau}{\mu}, \ x \in \mathbb{R}.$$

Since u is a weak solution of (3.1), the same does u_μ. We claim that

$$u_\mu \longrightarrow U, \qquad f(u_\mu) \longrightarrow f(U), \qquad \text{in } L^1_{loc}((0, \infty) \times \mathbb{R}), \text{ as } \mu \to 0. \tag{3.22}$$

Let $R > 0$ and $\mu < \frac{\tau}{R}$. Since

$$U(\mu t, \mu x) = U(t, x), \qquad t > 0, \ x \in \mathbb{R},$$

we get

$$\int_{-R}^{R} \int_{-R}^{R} |u_\mu(t, x) - U(t, x)| dt dx$$

$$= \frac{1}{\mu^2} \int_{-R\mu}^{R\mu} \int_{-R\mu}^{R\mu} |u(t + \tau, x + \xi) - U(t, x)| dt dx \longrightarrow 0,$$

namely

$$u_\mu \longrightarrow U, \qquad \text{in } L^1((-R, R) \times (-R, R)), \text{ as } \mu \to 0.$$

Therefore the Dominated Convergence Theorem gives (3.22). Finally, Theorem 3.4 and (3.22) implies that U is a entropy solution of (3.1). Then, the claim follows from Theorem 3.7. □

Example 3.1 The function

$$u(t, x) = \begin{cases} -\dfrac{2}{3}\left(t + \sqrt{3x + t^2}\right) & \text{if } 4x + t^2 > 0, \\ 0 & \text{if } 4x + t^2 < 0 \end{cases} \tag{3.23}$$

is an entropy solution of the Cauchy problem

$$\begin{cases} u_t + \left(\dfrac{u^2}{2}\right)_x = 0, & t > 0, \ x \in \mathbb{R}, \\ u(0, x) = \begin{cases} -\dfrac{2}{\sqrt{3}}\sqrt{x} & \text{if } x > 0, \\ 0 & \text{if } x < 0. \end{cases} \end{cases} \tag{3.24}$$

Introduce the notation

$$u_-(t, x) = 0, \quad u_+(t, x) = -\frac{2}{3}\left(t + \sqrt{3x + t^2}\right),$$

$$\lambda(t) = -\frac{t^2}{4}, \quad f(\xi) = \frac{\xi^2}{2}.$$

Since

$$\partial_x u_+(t, x) = -\frac{1}{\sqrt{3x + t^2}},$$

$$\partial_t u_+(t, x) = -\frac{2}{3}\left(1 + \frac{t}{\sqrt{3x + t^2}}\right),$$

$$u_+(t, x)\partial_x u_+(t, x) = \frac{2}{3}\left(\frac{t}{\sqrt{3x + t^2}} + 1\right),$$

then u_- and u_+ are classical solutions of the Burgers equation.

We have only to verify that (3.16) and (3.17) hold along the curve $x = \lambda(t)$. Since

$$u_-(t, \lambda(t)) = 0,$$

$$u_+(t, \lambda(t)) = -t \leq 0,$$

$$f(u_+(t, \lambda(t))) - f(u_-(t, \lambda(t))) - \lambda'(t)(u_+(t, \lambda(t)) - u_-(t, \lambda(t))) = 0,$$

the Rankine-Hugoniot Condition is satisfied and the jump is downward (note that f is convex).

3.6   Change of Coordinates

One of the features of the weak and entropy solutions is that they are not invariant under changes of coordinates. These ones transform smooth solutions in smooth solutions but in general they do not trasform weak/entropy solutions in weak/entropy solutions. Let us consider the following simple example based on the Burgers equation. We know that the shock

$$u(t, x) = \begin{cases} 1, & \text{if } x < t/2, \\ 0, & \text{if } x \geq t/2, \end{cases} \tag{3.25}$$

provides an entropy solution of the Riemann problem

$$\partial_t u + \partial_x\left(\frac{u^2}{2}\right) = 0, \qquad u(0, x) = \begin{cases} 1, & \text{if } x < 0, \\ 0, & \text{if } x \geq 0. \end{cases} \tag{3.26}$$

Consider the new unknown

$$v = u^3.$$

(3.25) and (3.26) become

$$v(t, x) = \begin{cases} 1, & \text{if } x < t/2, \\ 0 & \text{if } x \geq t/2, \end{cases} \tag{3.27}$$

and

$$\partial_t v + \partial_x \left(\frac{3}{4} v^{4/3} \right) = 0, \qquad v(0, x) = \begin{cases} 1, & \text{if } x < 0, \\ 0 & \text{if } x \geq 0. \end{cases} \tag{3.28}$$

respectively. Since v does not satisfies the Rankine-Hugoniot condition, it does not provide a weak solution of (3.28).

3.7 Uniqueness and Stability of Entropy Solutions

In this section, we prove the Kružkov Theorem [1]. It has three main consequences:

- the uniqueness of the entropy solutions,
- the L^1 Lipschitz continuity with respect to the initial conditions of the entropy solutions,
- the finite speed of propagation of the waves generated by conservation laws.

Theorem 3.9 (Kružkov [1]) *Let $u, v : [0, \infty) \times \mathbb{R} \to \mathbb{R}$ be two entropy solutions of (3.1). If*

$$u, v \in L^\infty((0, \infty) \times \mathbb{R}),$$

then

$$\int_{-R}^{R} |u(t_2, x) - v(t_2, x)| dx \leq \int_{-R-L(t_2-t_1)}^{R+L(t_2-t_1)} |u(t_1, x) - v(t_1, x)| dx, \tag{3.29}$$

for every $R > 0$ and almost every $0 \leq t_1 \leq t_2$, where

$$L = \sup_{(0,\infty) \times \mathbb{R}} (|f'(u)| + |f'(v)|). \tag{3.30}$$

A fundamental consequence of Kružkov Theorem is the following.

Corollary 3.1 (Uniqueness and Stability of Entropy Solutions) *Let* u, v :
$[0, \infty) \times \mathbb{R} \to \mathbb{R}$ *be two entropy solutions of* (3.1). *If*

$$u, v \in L^\infty((0, \infty) \times \mathbb{R}),$$

$$u(0, \cdot) - v(0, \cdot) \in L^1(\mathbb{R}) \ (or \ u(0, \cdot), v(0, \cdot) \in L^1(\mathbb{R})),$$

then

$$u(t, \cdot) - v(t, \cdot) \in L^1(\mathbb{R}) \ (or \ u(t, \cdot), v(t, \cdot) \in L^1(\mathbb{R})),$$

$$\|u(t, \cdot) - v(t, \cdot)\|_{L^1(\mathbb{R})} \le \|u(0, \cdot) - v(0, \cdot)\|_{L^1(\mathbb{R})},$$

(3.31)

for almost every $t \ge 0$. *In particular*

$$u(0, \cdot) = v(0, \cdot) \Longrightarrow u = v.$$

The proof of the Kružkov Theorem is based on the following lemma.

Lemma 3.2 (Doubling of Variables) *Let* $u, v : [0, \infty) \times \mathbb{R} \to \mathbb{R}$ *be two entropy solutions of* (3.1). *If*

$$u, v \in L^\infty((0, \infty) \times \mathbb{R}),$$

then

$$\partial_t |u - v| + \partial_x \left(\text{sign} \, (u - v) \, (f(u) - f(v)) \right) \le 0 \tag{3.32}$$

holds in the sense of distributions on $(0, \infty) \times \mathbb{R}$.

Proof The *idea of the proof* is the following. Subtract the equations for u and v

$$\partial_t (u - v) + \partial_x (f(u) - f(v)) = 0,$$

multiply by $\text{sign} \, (u - v)$

$$\text{sign} \, (u - v) \, \partial_t (u - v) + \text{sign} \, (u - v) \, \partial_x (f(u) - f(v)) = 0,$$

and use the chain rule

$$\partial_t |u - v| + \partial_x (\text{sign} \, (u - v) \, (f(u) - f(v))) = 0.$$

Unfortunately we do not have enough regularity on u and v therefore we are not able to apply the chain rule. We have to use a sharper argument [1].

Let $\varphi = \varphi(t, s, x, y)$ be a C^∞ nonnegative test function defined on $(0, \infty) \times (0, \infty) \times \mathbb{R} \times \mathbb{R}$. Since u and v are entropy solutions of (3.1), we have

$$\int_0^\infty \int_\mathbb{R} \Big(|u(t, x) - v(s, y)| \partial_t \varphi(t, s, x, y)$$

$$+ \, \text{sign} \, (u(t, x) - v(s, y)) \cdot$$

$$\cdot (f(u(t, x)) - f(v(s, y))) \partial_x \varphi(t, s, x, y) \Big) dt dx \geq 0,$$

$$\int_0^\infty \int_\mathbb{R} \Big(|v(s, y) - u(t, x)| \partial_s \varphi(t, s, x, y)$$

$$+ \, \text{sign} \, (v(s, y) - u(t, x)) \cdot$$

$$\cdot (f(v(s, y)) - f(u(t, x))) \partial_y \varphi(t, s, x, y) \Big) ds dy \geq 0,$$

and then

$$\int_0^\infty \int_0^\infty \int_\mathbb{R} \int_\mathbb{R} \Big(|u(t, x) - v(s, y)| (\partial_t \varphi + \partial_s \varphi)$$

$$+ \, \text{sign} \, (u(t, x) - v(s, y)) \, (f(u(t, x)) - f(v(s, y))) \cdot \qquad (3.33)$$

$$\cdot (\partial_x \varphi + \partial_y \varphi) \Big) dt ds dx dy \geq 0.$$

Let $\psi \in C^\infty((0, \infty) \times \mathbb{R})$ be a nonnegative test function and $\delta \in C^\infty(\mathbb{R})$ be such that

$$\delta \geq 0, \qquad \|\delta\|_{L^1(\mathbb{R})} = 1, \qquad \text{supp}(\delta) \subset [-1, 1].$$

Define

$$\delta_n(x) = n\delta(nx),$$

$$\varphi_n(t, s, x, y) = \psi \left(\frac{t+s}{2}, \frac{x+y}{2} \right) \delta_n \left(\frac{s-t}{2} \right) \delta_n \left(\frac{y-x}{2} \right). \qquad (3.34)$$

We use φ_n as test function in (3.33)

$$\int_0^\infty \int_0^\infty \int_\mathbb{R} \int_\mathbb{R} \delta_n \left(\frac{s-t}{2} \right) \delta_n \left(\frac{y-x}{2} \right) \cdot$$

$$\cdot \Big((|u(t, x) - v(s, y)| \partial_t \psi \left(\frac{t+s}{2}, \frac{x+y}{2} \right)$$

$$+ \operatorname{sign}(u(t, x) - v(s, y)) (f(u(t, x)) - f(v(s, y))) \cdot$$

$$\cdot \, \partial_x \psi \left(\frac{t+s}{2}, \frac{x+y}{2} \right) \right) dt ds dx dy \geq 0.$$

As $n \to \infty$ we get

$$\int_0^\infty \int_{\mathbb{R}} (|u - v| \partial_t \psi + \operatorname{sign}(u - v) (f(u) - f(v)) \partial_x \psi) \, dt dx \geq 0.$$

$\square$

Proof of Theorem 3.9 Let $R > 0$ and $0 \leq t_1 \leq t_2$. The *idea of the proof* is the following. Integrate both sides of (3.32) on the set

$$\Omega = \left\{ (t, x) \in [0, \infty) \times \mathbb{R}; t_1 \leq t \leq t_2, |x| \leq R + L(t_2 - t) \right\} \tag{3.35}$$

and use the divergence theorem and (3.30)

$$0 \geq \iint_\Omega (\partial_t |u - v| + \partial_x (\operatorname{sign}(u - v) (f(u) - f(v)))) \, dt dx$$

$$= \iint_\Omega \operatorname{div}_{(t,x)} \begin{pmatrix} |u - v| \\ \operatorname{sign}(u - v) (f(u) - f(v)) \end{pmatrix} dt dx$$

$$= \int_{-R}^{R} |u(t_2, x) - v(t_2, x)| dx - \int_{-R-L(t_2-t_1)}^{R+L(t_2-t_1)} |u(t_1, x) - v(t_1, x)| dx$$

$$+ \int_{t_1}^{t_2} (L|u - v| + \operatorname{sign}(u - v) (f(u) - f(v))(t, R + L(t_2 - t))) \, dt$$

$$- \int_{t_1}^{t_2} (L|u - v| + \operatorname{sign}(u - v) (f(u) - f(v))(t, -R - L(t_2 - t))) \, dt$$

$$\geq \int_{-R}^{R} |u(t_2, x) - v(t_2, x)| dx - \int_{-R-L(t_2-t_1)}^{R+L(t_2-t_1)} |u(t_1, x) - v(t_1, x)| dx.$$

Unfortunately we do not have enough regularity on u and v therefore we are not able to apply the chain rule. We have to use a sharper argument [1].

Define

$$\alpha_n(x) = \int_{-\infty}^{x} \delta_n(y) dy, \qquad x \in \mathbb{R},$$

where δ_n is defined in (3.34). Consider the test function

$$\varphi_n(t,x) = (\alpha_n(t-t_1) - \alpha_n(t-t_2))\left(1 - \alpha_n\left(\sqrt{x^2 + \frac{1}{n}} - R - L(t_2 - t)\right)\right),$$

that is a smooth approximant of the characteristic function of the set Ω defined in (3.35). Testing (3.32) with φ_n we get

$$\int_0^\infty \int_{\mathbb{R}} |u - v| \, (\delta_n(t-t_1) - \delta_n(t-t_2)) \cdot$$

$$\cdot \left(1 - \alpha_n\left(\sqrt{x^2 + \frac{1}{n}} - R - L(t_2 - t)\right)\right) dt dx$$

$$- L \int_0^\infty \int_{\mathbb{R}} |u - v| \, (\alpha_n(t-t_1) - \alpha_n(t-t_2)) \cdot$$

$$\cdot \delta_n\left(\sqrt{x^2 + \frac{1}{n}} - R - L(t_2 - t)\right) dt dx$$

$$+ \int_0^\infty \int_{\mathbb{R}} \operatorname{sign}(u - v) \, (f(u) - f(v)) \, (\alpha_n(t-t_1) - \alpha_n(t-t_2)) \cdot$$

$$\cdot \frac{x}{\sqrt{x^2 + \frac{1}{n}}} \delta_n\left(\sqrt{x^2 + \frac{1}{n}} - R - L(t_2 - t)\right) dt dx \geq 0.$$

Since

$$|f(u) - f(v)| \leq |u - v|, \qquad \left| \frac{x}{\sqrt{x^2 + \frac{1}{n}}} \right| \leq 1,$$

we have

$$\int_0^\infty \int_{\mathbb{R}} |u - v| \, (\delta_n(t-t_1) - \delta_n(t-t_2)) \cdot$$

$$\cdot \left(1 - \alpha_n\left(\sqrt{x^2 + \frac{1}{n}} - R - L(t_2 - t)\right)\right) dt dx$$

$$\geq \int_0^\infty \int_{\mathbb{R}} \left(L|u - v| - \operatorname{sign}(u - v)(f(u) - f(v))\frac{x}{\sqrt{x^2 + \frac{1}{n}}}\right) \cdot$$

$$\cdot (\alpha_n(t-t_1) - \alpha_n(t-t_2))\delta_n\left(\sqrt{x^2 + \frac{1}{n}} - R - L(t_2 - t)\right) dt dx \geq 0.$$

As $n \to \infty$, using the fact that, due to the Lusin Theorem, the map $t \geq 0 \mapsto u(t, \cdot) - v(t, \cdot) \in L^1_{loc}(\mathbb{R})$ is almost everywhere continuous, we get (3.29).   □

3.8 Exercises

Exercise 3.1 Check that the function

$$u(t, x) = \begin{cases} 0, & \text{if } x < t/2, \\ 1, & \text{if } x \geq t/2, \end{cases} \qquad (3.36)$$

provides a weak solution of the Riemann problem

$$\partial_t u + \partial_x \left(\frac{u^2}{2} \right) = 0, \qquad u(0, x) = \begin{cases} 0, & \text{if } x < 0, \\ 1, & \text{if } x \geq 0. \end{cases} \qquad (3.37)$$

Exercise 3.2 Check that the function

$$u(t, x) = \begin{cases} 0, & \text{if } x < t/2, \\ 1, & \text{if } x \geq t/2, \end{cases} \qquad (3.38)$$

does not provide a weak solution of the Riemann problem

$$\partial_t u + \partial_x \left(\frac{4}{5} u^{5/4} \right) = 0, \qquad u(0, x) = \begin{cases} 1, & \text{if } x < 0, \\ 0, & \text{if } x \geq 0. \end{cases} \qquad (3.39)$$

Exercise 3.3 Solve the following Riemann problems

$$\begin{cases} u_t + (\ln(u))_x = 0, & t > 0, \ x \in \mathbb{R}, \\ u(0, x) = \begin{cases} e, & \text{if } x \geq 0, \\ 1, & \text{if } x < 0, \end{cases} \\ u_t + (\ln(u))_x = 0, & t > 0, \ x \in \mathbb{R}, \\ u(0, x) = \begin{cases} 1, & \text{if } x \geq 0, \\ e, & \text{if } x < 0. \end{cases} \end{cases} \qquad (3.40)$$

Exercise 3.4 Solve the following Riemann problems

$$
\begin{cases}
u_t + \left(u^4 - u^2\right)_x = 0, & t > 0,\ x \in \mathbb{R}, \\[2mm]
u(0, x) = \begin{cases} -2, & \text{if } x \geq 0, \\ 3, & \text{if } x < 0. \end{cases}
\end{cases}
$$

$$
\begin{cases}
u_t + \left(u^4 - u^2\right)_x = 0, & t > 0,\ x \in \mathbb{R}, \\[2mm]
u(0, x) = \begin{cases} -3, & \text{if } x \geq 0, \\ 2, & \text{if } x < 0. \end{cases}
\end{cases}
\tag{3.41}
$$

References

1. Kružkov, S.N.: First order quasilinear equations with several independent variables. Mat. Sb. (N.S.) **81**(123), 228–255 (1970)
2. Liu, T.-P., Ruggeri, T.: Entropy production and admissibility of shocks. Acta Math. Appl. Sin. Engl. Ser. **19**(1), 1–12 (2003)

Chapter 4
Riemann Problem

Abstract This chapter is dedicated to the explicit entropy solution of the Riemann problem, that is a Cauchy problem with an initial datum of Heaviside type.

Keywords Riemann problem · Shock · Rarefaction

In Chap. 3 we proved the uniqueness and stability of entropy solutions of Cauchy problems. Here we focus on the existence of entropy solutions. We analyze the simplest possible case: the Riemann problems, that are Cauchy problems with Heaviside type initial condition

$$\begin{cases} \partial_t u + \partial_x f(u) = 0, & t > 0,\ x \in \mathbb{R}, \\ u(0, x) = \begin{cases} u_+, & \text{if } x \geq 0, \\ u_-, & \text{if } x < 0, \end{cases} \end{cases} \tag{4.1}$$

where $f \in C^2(\mathbb{R})$ and $u_- \neq u_+$ are constants.

Riemann introduced (4.1) in his paper on ideal gases [1].

In the following sections we first consider the case in which f is convex. Indeed the solutions obtained under that assumption are the building blocks of the solutions of the general case.

4.1 Strictly Convex Fluxes

We assume that f is convex, the concave case is analogous.

We distinguish two cases. If (see Fig. 4.1)

$$u_+ < u_-,$$

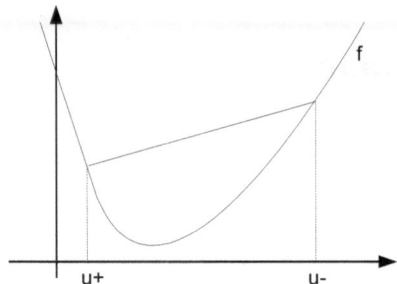

Fig. 4.1 Riemann problem with $u_- > u_+$ and f convex

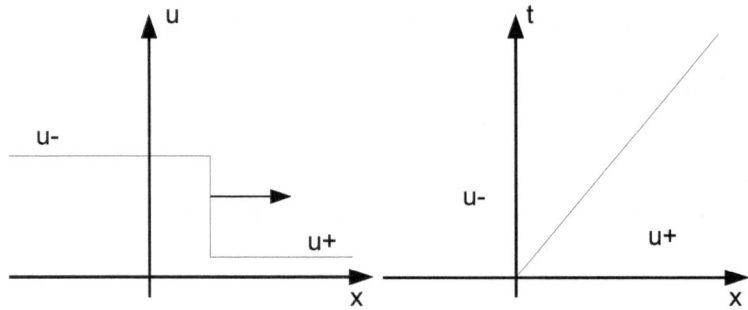

Fig. 4.2 Shock with $u_- > u_+$ and f convex

the entropy solution of (4.1) is the *shock* (see Fig. 4.2)

$$u(t, x) = \begin{cases} u_+, & \text{if } x \geq \dfrac{f(u_+) - f(u_-)}{u_+ - u_-}t, \\ u_-, & \text{if } x < \dfrac{f(u_+) - f(u_-)}{u_+ - u_-}t. \end{cases}$$

If (see Fig. 4.3)

$$u_+ > u_-,$$

the entropy solution of (4.1) is the *rarefaction* (see Fig. 4.4)

$$u(t, x) = \begin{cases} u_+, & \text{if } x \geq f'(u_+)t, \\ \sigma, & \text{if } x = f'(\sigma)t, \ u_- \leq \sigma < u_+, \\ u_-, & \text{if } x = f'(u_-)t. \end{cases} \tag{4.2}$$

Observe that the definition makes sense because f is convex and then f' is increasing.

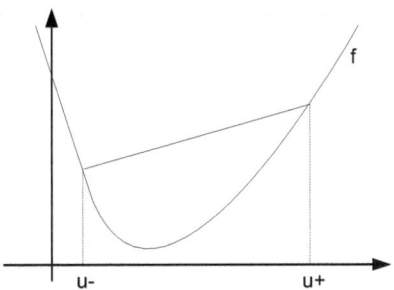

Fig. 4.3 Riemann problem with $u_- < u_+$ and f convex

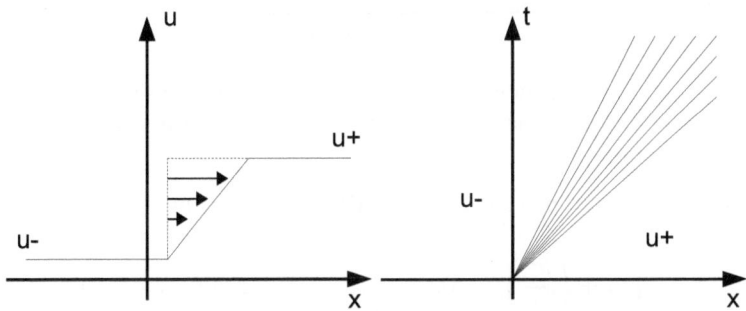

Fig. 4.4 Rarefaction wave with $u_- < u_+$ and f convex

We claim that

$$\partial_t \eta(u) + \partial_x q(u) = 0, \tag{4.3}$$

for every entropy η with flux q, where u is the rarefaction wave defined in (4.2).
    Consider the sets

$$\begin{aligned}
\Omega_1 &= \{(t, x) \in (0, \infty) \times \mathbb{R}; \, x \leq f'(u_-)t\}, \\
\Omega_2 &= \{(t, x) \in (0, \infty) \times \mathbb{R}; \, f'(u_-)t < x < f'(u_+)t\}, \\
\Omega_3 &= \{(t, x) \in (0, \infty) \times \mathbb{R}; \, x \geq f'(u_+)t\},
\end{aligned}$$

with outer normals n_1, n_2, n_3, and a nonnegative test function $\varphi \in C^\infty((0, \infty) \times \mathbb{R})$. We have

$$\int_0^\infty \int_\mathbb{R} (\eta(u)\partial_t \varphi + q(u)\partial_x \varphi)dtdx$$

$$= -\sum_{i=1}^3 \int\!\!\int_{\Omega_i} \underbrace{(\partial_t \eta(u) + \partial_x q(u))}_{=0 \text{ (because } u \text{ is smooth in each } \Omega_i)} \varphi dtdx$$

$$+ \sum_{i=1}^3 \int_{\partial\Omega_i} (\eta(u)\varphi, q(u)\varphi) \cdot n_i d\sigma = 0.$$

$$\underbrace{\phantom{+ \sum_{i=1}^3 \int_{\partial\Omega_i} (\eta(u)\varphi, q(u)\varphi) \cdot n_i d\sigma = 0.}}_{=0 \text{ (because } u \text{ is continuous)}}$$

Therefore (4.3) holds and then (4.2) is the entropy solution of (4.1).

When f is concave we have a completely symmetric case, a shock when $u_- < u_+$ and a rarefaction when $u_- > u_+$.

Example 4.1 The entropy solution of the Riemann problem

$$\begin{cases} u_t + \left(\dfrac{u^2}{2}\right)_x = 0, & t > 0, \ x \in \mathbb{R}, \\ u(0, x) = \begin{cases} -1, & \text{if } x < 0, \\ 1, & \text{if } x \geq 0, \end{cases} \end{cases}$$

is the rarefaction wave

$$u(t, x) = \begin{cases} -1, & \text{if } x < -t, \\ \sigma, & \text{if } x = \sigma t, \ -1 < \sigma \leq 1, \\ 1, & \text{if } x > t. \end{cases}$$

Example 4.2 The entropy solution of the Riemann problem

$$\begin{cases} u_t + (u^3)_x = 0, & t > 0, \ x \in \mathbb{R}, \\ u(0, x) = \begin{cases} 1, & \text{if } x < 2, \\ 0, & \text{if } x > 2, \end{cases} \end{cases}$$

is the shock

$$u(t, x) = \begin{cases} 1, & \text{if } x < t + 2, \\ 0, & \text{if } x \geq t + 2. \end{cases}$$

Example 4.3 The entropy solution of the Riemann problem

$$
\begin{cases}
u_t + (u^3)_x = 0, & t > 0,\ x \in \mathbb{R}, \\[4pt]
u(0, x) = \begin{cases} 0, & \text{if } x < 2, \\ 1, & \text{if } x \geq 2, \end{cases} &
\end{cases}
$$

is the rarefaction wave

$$
u(t, x) = \begin{cases}
0, & \text{if } x < 2, \\
\sigma, & \text{if } x = 3\sigma^2 t + 2,\ 0 < \sigma \leq 1, \\
1, & \text{if } x > 3t + 2.
\end{cases}
$$

Example 4.4 The entropy solution of the Riemann problem

$$
\begin{cases}
u_t + (e^u)_x = 0, & t > 0,\ x \in \mathbb{R}, \\[4pt]
u(0, x) = \begin{cases} 2, & \text{if } x < 0, \\ 0, & \text{if } x > 0, \end{cases} &
\end{cases}
$$

is the shock

$$
u(t, x) = \begin{cases}
1, & \text{if } x < (e^2 - 1)t/2, \\
0, & \text{if } x \geq (e^2 - 1)t/2.
\end{cases}
$$

4.2 General Fluxes

In the case of convex or concave fluxes the solution of the Riemann problem (4.1) consists of only one wave, a shock or a rarefaction. In the case of fluxes that are not convex or concave we can have several waves of both types. Moreover, the waves may also be glued together.

We have to distinguish again two cases. If

$$
u_- < u_+,
$$

we consider the convex hull f_* of f in the interval $[u_-, u_+]$, i.e., f_* is the largest convex map such that

$$
f_*(\xi) \leq f(\xi), \qquad u_- \leq \xi \leq u_+.
$$

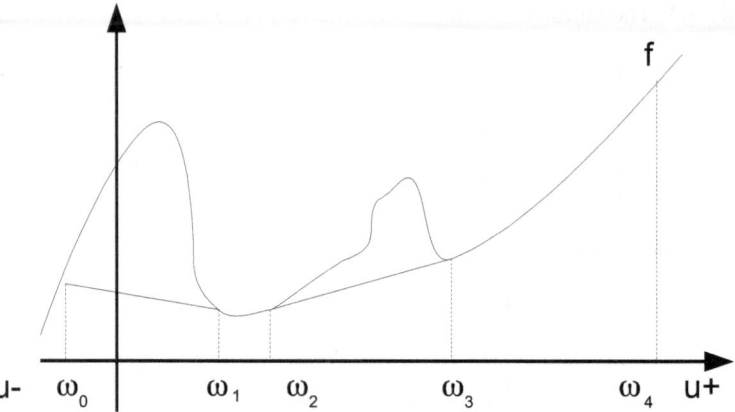

Fig. 4.5 Riemann problem with $u_- < u_+$

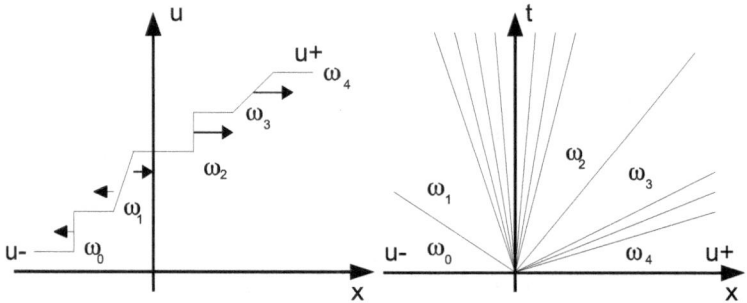

Fig. 4.6 Solution of the Riemann problem with $u_- < u_+$

Let consider the points $w_0, \ldots, w_n$ such that (see Fig. 4.5)

$$u_- = w_0 < w_1 < \ldots < w_n = u_+,$$

$$f(w_i) = f_*(w_i), \quad i = 0, \ldots, n,$$

$$w_i < u < w_{i+1} \Rightarrow f_*(u) < f(u) \text{ or } f_*(u) = f(u), \quad i = 0, \ldots, n-1.$$

We solve separately the $n-1$ Riemann problems obtained in correspondence of
the values (w_i, w_{i+1}), $i = 0, \ldots, n-1$. If $f < f_*$ in (w_i, w_{i+1}), we have a shock;
otherwise a rarefaction (see Fig. 4.6). This algorithm provides clearly the entropy
solution of (4.1) because we are gluing entropy solutions.
   If

$$u_- > u_+,$$

Fig. 4.7 The Riemann
problem (4.4)

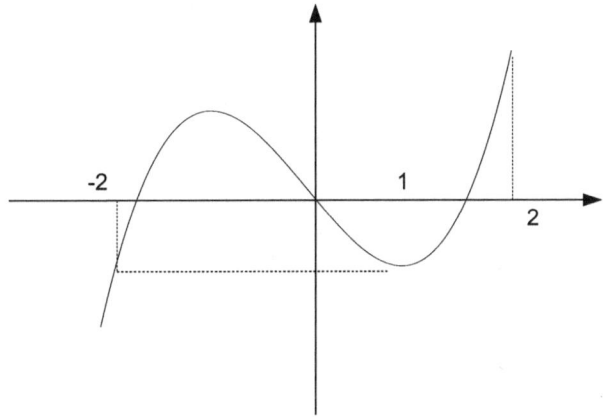

we consider the concave hull f^* of f in the interval $[u_-, u_+]$, i.e., f^* is the smallest
concave map such that

$$f(\xi) \le f^*(\xi), \qquad u_- \le \xi \le u_+,$$

and we argue in the same way.

Example 4.5   Consider the Riemann problem (see Fig. 4.7)

$$\begin{cases} \partial_t u + \partial_x (u^3 - 3u) = 0, & t > 0,\ x \in \mathbb{R}, \\ u(0, x) = \begin{cases} 2, & \text{if } x \ge 0, \\ -2, & \text{if } x < 0, \end{cases} \end{cases} \qquad (4.4)$$

The solution of (4.4) is (see Fig. 4.8)

$$u(t, x) = \begin{cases} 2, & \text{if } x \ge 9t, \\ \sigma, & \text{if } x = (3\sigma^2 - 3)t,\ 1 \le \sigma < 2, \\ -2, & \text{if } x < 0, \end{cases}$$

where the shock connecting -2 and 1 is attached to the rarefaction from 1 to 2. The
same feature can be found in

$$\begin{cases} \partial_t u + \partial_x (u^3 - 3u) = 0, & t > 0,\ x \in \mathbb{R}, \\ u(0, x) = \begin{cases} -2, & \text{if } x \ge 0, \\ 2, & \text{if } x < 0. \end{cases} \end{cases}$$

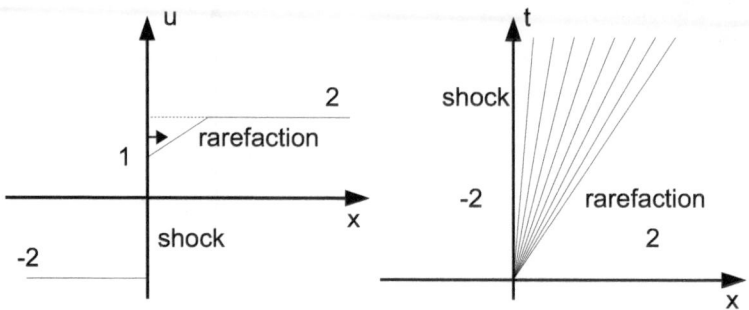

Fig. 4.8 Solution of the Riemann problem (4.4)

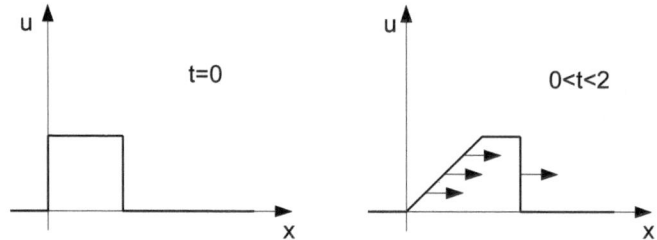

Fig. 4.9 Solution of the Cauchy problem (4.5) for $0 \leq t < 2$

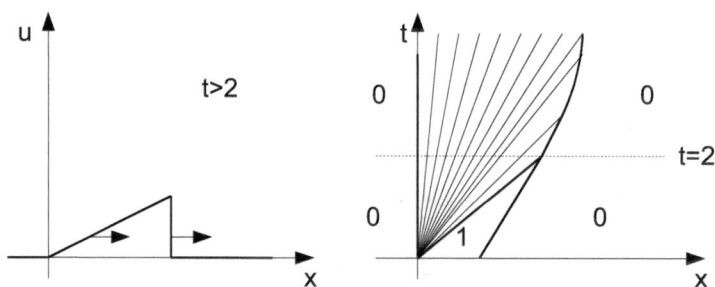

Fig. 4.10 Solution of the Cauchy problem (4.5) for $t \geq 2$

Example 4.6 Let us solve the Cauchy problem (Figs. 4.9 and 4.10)

$$\begin{cases} u_t + \left(\dfrac{u^2}{2} \right)_x = 0, & t > 0, \ x \in \mathbb{R}, \\[2mm] u(0, x) = \begin{cases} 1, & \text{if } 0 < x < 1, \\ 0, & \text{otherwise.} \end{cases} \end{cases} \qquad (4.5)$$

The wave generated at $x = 0$ is a rarefaction wave with speeds between 0 and 1, the one generated at $x = 1$ a shock with speed 1/2, they interact at $t = 2$, and we have

$$u(t, x) = \begin{cases} 0, & \text{if } x \leq 0, \\ \sigma, & \text{if } x = \sigma t, \ 0 \leq \sigma \leq 1, \\ 1, & \text{if } t < x \leq t/2 + 1, \\ 0, & \text{if } x > t/2 + 1, \end{cases} \qquad 0 \leq t \leq 2. \qquad (4.6)$$

For $t \geq 2$, we have a structure of the type

$$u(t, x) = \begin{cases} 0, & \text{if } x \leq 0, \\ \sigma, & \text{if } x = \sigma t, \ 0 \leq \sigma \leq \lambda(t)/t, \qquad t \geq 2. \\ 0, & \text{if } x > \lambda(t), \end{cases} \qquad (4.7)$$

We have to determine $\lambda(t)$. We know that

$$\lambda(2) = 2. \qquad (4.8)$$

The Rankine-Hugoniot condition gives

$$\lambda'(t) = \frac{u(t, \lambda(t)^-)}{2}. \qquad (4.9)$$

Finally, from (4.7), we know

$$u(t, \lambda(t)^-) = \frac{\lambda(t)}{t}. \qquad (4.10)$$

Therefore, (4.8), (4.9), and (4.10) imply that $\lambda(t)$ is the unique solution of the ordinary differential problem

$$\lambda'(t) = \frac{\lambda(t)}{2t}, \qquad \lambda(2) = 2,$$

namely

$$\lambda(t) = \sqrt{2t}, \qquad t \geq 2.$$

Example 4.7 Let us solve the Cauchy problem

$$
\begin{cases}
u_t + \left(\dfrac{u^2}{2}\right)_x = 0, & t > 0,\ x \in \mathbb{R}, \\[2mm]
u(0, x) = \begin{cases}
1, & \text{if } x < -1, \\
0, & \text{if } -1 < x < 0, \\
2, & \text{if } 0 < x < 1, \\
0, & \text{if } x > 1.
\end{cases}
\end{cases}
\tag{4.11}
$$

The wave generated at $x = -1$ is a shock with speed 1/2, the one generated at $x = 0$ is a rarefaction wave with speeds between 0 and 2, the one generated at $x = 1$ a shock with speed 1. The first interaction is between the second and the third wave at $t = 1$, and we have

$$
u(t, x) = \begin{cases}
1, & \text{if } x \le t/2 - 1, \\
0, & \text{if } t/2 - 1 \le x \le 0, \\
\sigma, & \text{if } x = \sigma t,\ 0 \le \sigma \le 2, \qquad 0 \le t \le 1. \\
2, & \text{if } 2t < x \le t + 1, \\
0, & \text{if } x > t + 1,
\end{cases}
\tag{4.12}
$$

The second interaction is between the first and the second wave at $t = 2$, and for $1 \le t \le 2$ and $t \ge 2$ we have a structure of the type

$$
u(t, x) = \begin{cases}
1, & \text{if } x \le 0, \\
\sigma, & \text{if } x = \sigma t,\ 0 \le \sigma \le \lambda(t)/t, \qquad 1 \le t \le 2, \\
0, & \text{if } x > \lambda(t),
\end{cases}
\tag{4.13}
$$

$$
u(t, x) = \begin{cases}
1, & \text{if } x \le \gamma(t), \\
\sigma, & \text{if } x = \sigma t,\ \gamma(t)/t \le \sigma \le \lambda(t)/t, \qquad t \ge 2. \\
0, & \text{if } x > \lambda(t),
\end{cases}
\tag{4.14}
$$

We have to determine $\gamma(t)$ and $\lambda(t)$ (Figs. 4.11, 4.12, and 4.13). We know that

$$
\gamma(2) = 0, \qquad \lambda(1) = 2.
\tag{4.15}
$$

The Rankine-Hugoniot condition gives

$$
\gamma'(t) = \frac{1 + u(t, \gamma(t)^+)}{2}, \qquad \lambda'(t) = \frac{u(t, \lambda(t)^-)}{2}.
\tag{4.16}
$$

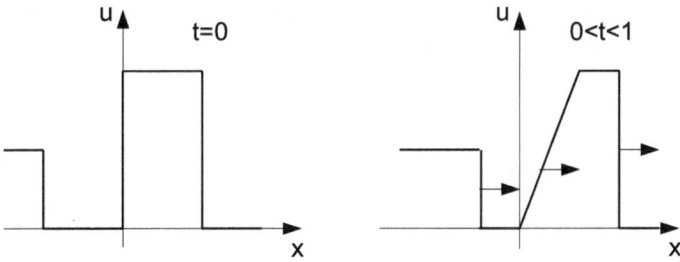

Fig. 4.11 Solution of the Cauchy problem (4.11) for $0 \le t < 1$

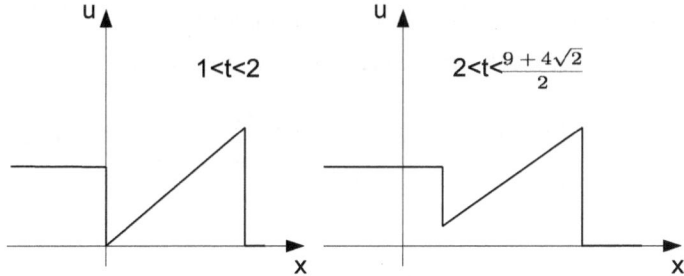

Fig. 4.12 Solution of the Cauchy problem (4.11) for $1 \le t < 2$ and $2 \le t < \frac{9+4\sqrt{2}}{2}$

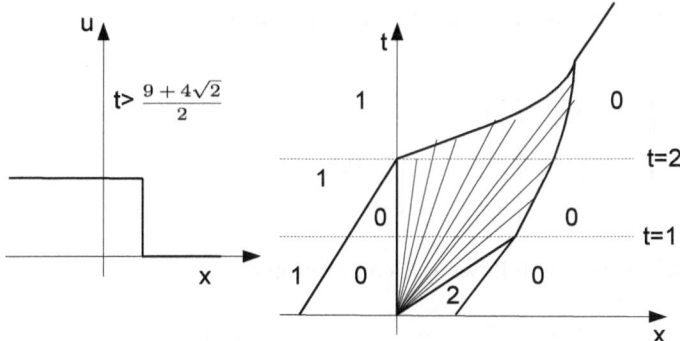

Fig. 4.13 Solution of the Cauchy problem (4.11) for $t \ge \frac{9+4\sqrt{2}}{2}$

Finally, from (4.13), we know

$$u(t, \gamma(t)^+) = \frac{\gamma(t)}{t}, \qquad u(t, \lambda(t)^-) = \frac{\lambda(t)}{t}. \tag{4.17}$$

Therefore, (4.8), (4.9), and (4.10) imply that $\gamma(t)$ and $\lambda(t)$ are the unique solution of the ordinary differential problems

$$\begin{cases} \gamma'(t) = \dfrac{1}{2}\left(1 + \dfrac{\gamma(t)}{t}\right), \\ \gamma(2) = 0, \end{cases} \qquad \begin{cases} \lambda'(t) = \dfrac{\lambda(t)}{2t} \\ \lambda(1) = 2, \end{cases}$$

namely

$$\gamma(t) = t - \sqrt{2t}, \qquad \lambda(t) = 2\sqrt{t}.$$

Since, γ and λ interact at $\frac{9+4\sqrt{2}}{2}$, (4.14) holds only for $2 \le t \le \frac{9+4\sqrt{2}}{2}$. For $t \ge \frac{9+4\sqrt{2}}{2}$ we have only a shock connecting 0 and 1 with speed $\frac{1}{2}$

$$u(t, x) = \begin{cases} 1, & \text{if } x \le t/2 + \sqrt{18 + 8\sqrt{2}}, \\ 0, & \text{if } x > t/2 + \sqrt{18 + 8\sqrt{2}}, \end{cases} \qquad t \ge \frac{9 + 4\sqrt{2}}{2}.$$

4.3   Exercises

Exercise 4.1 Solve the following Riemann problem

$$\begin{cases} u_t + \left(\dfrac{u^2}{u^2 + (1-u)^2}\right)_x = 0, & t > 0, \ x \in \mathbb{R}, \\[2mm] u(0, x) = \begin{cases} 1, & \text{if } x \ge 0, \\ 0, & \text{if } x < 0. \end{cases} \end{cases} \tag{4.18}$$

Exercise 4.2 Solve the following Riemann problem

$$\begin{cases} u_t + \left(u^4 - u^2\right)_x = 0, & t > 0, \ x \in \mathbb{R}, \\[2mm] u(0, x) = \begin{cases} -1, & \text{if } x \ge 0, \\ 2, & \text{if } x < 0. \end{cases} \end{cases} \tag{4.19}$$

Exercise 4.3 Solve the following Riemann problem

$$\begin{cases} u_t + \left(u^4 - u^2\right)_x = 0, & t > 0, \ x \in \mathbb{R}, \\[2mm] u(0, x) = \begin{cases} 1, & \text{if } x \ge 0, \\ -2, & \text{if } x < 0. \end{cases} \end{cases} \tag{4.20}$$

Exercise 4.4 Solve the following Riemann problem

$$
\begin{cases}
u_t + (u^3)_x = 0, & t > 0, \ x \in \mathbb{R}, \\
u(0, x) = \begin{cases} 1, & \text{if } x \geq 0, \\ -1, & \text{if } x < 0. \end{cases}
\end{cases}
\tag{4.21}
$$

Exercise 4.5 Solve the following Riemann problem

$$
\begin{cases}
u_t + \left(\dfrac{1}{1 + u^2} \right)_x = 0, & t > 0, \ x \in \mathbb{R}, \\
u(0, x) = \begin{cases} 1/\sqrt{3}, & \text{if } x \geq 0, \\ 0, & \text{if } x < 0. \end{cases}
\end{cases}
\tag{4.22}
$$

Exercise 4.6 Solve the following Riemann problem

$$
\begin{cases}
u_t + \left(\dfrac{1}{1 + u^2} \right)_x = 0, & t > 0, \ x \in \mathbb{R}, \\
u(0, x) = \begin{cases} 1/\sqrt{3}, & \text{if } x \geq 0, \\ 1, & \text{if } x < 0. \end{cases}
\end{cases}
\tag{4.23}
$$

Exercise 4.7 Solve the following Cauchy problem

$$
\begin{cases}
u_t + (e^u)_x = 0, & t > 0, \ x \in \mathbb{R}, \\
u(0, x) = \begin{cases} -1, & \text{if } x \geq 1, \\ 0, & \text{if } -1 \leq x < 1, \\ 1, & \text{if } x < -1. \end{cases}
\end{cases}
\tag{4.24}
$$

Exercise 4.8 Solve the following Cauchy problem

$$
\begin{cases}
u_t + (e^u)_x = 0, & t > 0, \ x \in \mathbb{R}, \\
u(0, x) = \begin{cases} 1, & \text{if } x \geq 1, \\ 0, & \text{if } -1 \leq x < 1, \\ -1, & \text{if } x < -1. \end{cases}
\end{cases}
\tag{4.25}
$$

Exercise 4.9 Solve the following Cauchy problem

$$
\begin{cases}
u_t + (u^3)_x = 0, & t > 0, \ x \in \mathbb{R}, \\[2mm]
u(0, x) = \begin{cases}
-1, & \text{if } x \geq 1, \\
0, & \text{if } -1 \leq x < 1, \\
1, & \text{if } x < -1.
\end{cases}
\end{cases}
\tag{4.26}
$$

Reference

1. Riemann, B.: Über die Fortpflanzung ebener Luftwellen von endlicher Schwingungsweite. Dieterich (1860)

Chapter 5
Functions with Bounded Variation

Abstract This chapter is dedicated to a review of the space of functions with bounded variation in 1 and 2 dimensions. Indeed, BV is the space in which entropy solutions live if the initial data belong to BV. We give several examples and prove some qualitative results and some compactness theorems that play a key role in the existence results. The proof and the definitions require only the knowledge of some basic measure theory because we do not consider the general n-dimensional case.

Keywords Functions with bounded variation · Total variation · Absolutely continuous functions · Helly Theorem · Compactness

In this chapter we review the space of functions with bounded variation in 1 and 2 dimensions. BV is a space of possibly discontinuous functions in which the solutions of conservation laws naturally live. The complete and general theory of functions with bounded variation can be found in [1].

5.1 Definition and Structural Properties

Let $u : \mathbb{R} \to \mathbb{R}$ be a function. We call *total variation* of u the quantity [2]

$$TV(u) = \sup_{x_0 < \dots < x_N} \sum_{j=1}^{N} \left| u(x_j) - u(x_{j-1}) \right|.$$

If

$$TV(u) < \infty$$

we say that u has *bounded variation* and we write

$$u \in BV(\mathbb{R}).$$

G. M. Coclite, *Scalar Conservation Laws*, SpringerBriefs in Mathematics,
https://doi.org/10.1007/978-981-97-3984-4_5

Example 5.1 (Heaviside Function) The Heaviside function

$$u : \mathbb{R} \longrightarrow \mathbb{R}, \qquad u(x) = \begin{cases} 1, & \text{if } x \geq 0, \\ 0, & \text{if } x < 0, \end{cases}$$

has bounded variation. We have

$$u \in BV(\mathbb{R}), \qquad TV(u) = 1.$$

Example 5.2 (Piecewise Constant Functions) Let $\alpha_0, \ldots, \alpha_{N+1}, x_0, \ldots, x_N \in \mathbb{R}$ such that

$$x_0 < \ldots < x_N,$$

and consider the piecewise constant function

$$u = \alpha_0 \chi_{(-\infty, x_0)} + \sum_{j=1}^{N} \alpha_j \chi_{[x_{j-1}, x_j)} + \alpha_{N+1} \chi_{[x_N, \infty)}.$$

We have

$$u \in BV(\mathbb{R}), \qquad TV(u) = \sum_{j=0}^{N} \left| \alpha_j - \alpha_{j+1} \right|.$$

Example 5.3 (Monotone Bounded Functions) Let $u : \mathbb{R} \longrightarrow \mathbb{R}$ be a function. If u is monotone and bounded, then

$$u \in BV(\mathbb{R}), \qquad TV(u) = |u(\infty) - u(-\infty)|.$$

Example 5.4 (Absolutely Continuous Functions) Let $u : \mathbb{R} \longrightarrow \mathbb{R}$ be a function with distributional derivative u'. If

$$u' \in L^1(\mathbb{R}),$$

then

$$u \in BV(\mathbb{R}), \qquad TV(u) = \left\| u' \right\|_{L^1(\mathbb{R})}.$$

In particular, all the absolutely continuous functions have bounded variation, i.e.,

$$W^{1,1}(\mathbb{R}) \subset BV(\mathbb{R}).$$

Theorem 5.1 *Let* $u : \mathbb{R} \longrightarrow \mathbb{R}$ *be a function. If*

$$u \in BV(\mathbb{R}),$$

then

$$u \in L^{\infty}(\mathbb{R}),$$

and

$$\forall x \in \mathbb{R} \cup \{\pm\infty\} : \|u\|_{L^{\infty}(\mathbb{R})} \leq |u(x)| + TV(u). \tag{5.1}$$

Proof Let $x \in \mathbb{R}$, $x_0 \in \mathbb{R} \cup \{\pm\infty\}$. We have to prove that

$$|u(x)| \leq |u(x_0)| + TV(u). \tag{5.2}$$

We distinguish three cases.
If $x = x_0$, (5.2) trivially holds.
If $x \neq x_0$ and $x_0 \in \mathbb{R}$, we have

$$|u(x)| \leq |u(x_0)| + |u(x) - u(x_0)|$$

$$\leq |u(x_0)| + \sup_{x_0 < \dots < x_N} \sum_{j=1}^{N} |u(x_j) - u(x_{j-1})|$$

$$= |u(x_0)| + TV(u),$$

that is (5.2).
If $x_0 \in \{\pm\infty\}$, (5.2) follows from the previous case and the definition of limit.

$\square$

Theorem 5.2 *Let* $u : \mathbb{R} \longrightarrow \mathbb{R}$ *be a function. If*

$$u \in BV(\mathbb{R}),$$

then

(i) *for every* $x_0 \in R$ *there exist* $\lim\limits_{x \to x_0^+} u(x) \in R$ *and* $\lim\limits_{x \to x_0^-} u(x) \in R$;
(ii) *the limits* $\lim\limits_{x \to \infty} u(x) \in R$ *and* $\lim\limits_{x \to \infty} u(x) \in R$ *exist;*
(iii) *u has at most countably many points of discontinuity.*

Proof We begin by proving (i). Let $x_0 \in \mathbb{R}$ and $\{x_n\}_{n \in \mathbb{N}}$ be an increasing sequence such that $x_n \to x_0$. We have

$$\sum_{n=1}^{\infty} |u(x_n) - u(x_{n-1})| \leq TV(u) < \infty,$$

as a consequence $\{u(x_n)\}_{n \in \mathbb{N}}$ is a Cauchy sequence. Since $\{u(x_n)\}_{n \in \mathbb{N}}$ converges, we define

$$u(x_0^-) = \lim_n u(x_n).$$

We have to prove that the definition of $u(x_0^-)$ does not depend on the choice of the sequence $\{x_n\}_{n \in \mathbb{N}}$. Let $\{x_n'\}_{n \in \mathbb{N}}$ be one more increasing sequence such that $x_n' \to x_0$ and consider the set

$$A = \{x_n; n \in \mathbb{N}\} \cup \{x_n'; n \in \mathbb{N}\}.$$

Since A is countable and accumulates only in x_0 there exists an increasing sequence $\{x_n''\}_{n \in \mathbb{N}}$ such that

$$A = \{x_n''; n \in \mathbb{N}\}, \qquad x_n'' \longrightarrow x_0.$$

We have

$$\sum_{n=1}^{\infty} |u(x_n'') - u(x_{n-1}'')| \leq TV(u) < \infty,$$

as a consequence $\{u(x_n'')\}_{n \in \mathbb{N}}$ is a Cauchy sequence. Therefore, we have

$$\lim_n u(x_n') = \lim_n u(x_n'') = \lim_n u(x_n) = u(x_0^-).$$

The same argument works for $u(x_0^+)$ and (ii). We have to prove (iii). For every $n \in \mathbb{N} \setminus \{0\}$, we define the set

$$A_n = \left\{ x \in \mathbb{R}; \left| u(x^-) - u(x) \right| + \left| u(x^+) - u(x) \right| > \frac{1}{n} \right\}.$$

From the definition of total variation we get that A_n has finitely many points and

$$\#(A_n) \leq nTV(u).$$

Since

$$\bigcup_n A_n = \{x \in \mathbb{R}; u \text{ discontinuous in } x\},$$

(iii) is proved.   □

Remark 5.1 Thanks to the previous theorem a function u with bounded variation can be defined in every discontinuity point x as $u(x) = u(x^+)$. In this way, we get a right continuous function with the same total variation.

Theorem 5.3 *Let $u : \mathbb{R} \to \mathbb{R}$ and $\varepsilon > 0$. If*

$$u \text{ is right continuous,} \qquad u \in BV(\mathbb{R}),$$

then there exists a function $v : \mathbb{R} \to \mathbb{R}$ such that

$$v \text{ is piecewise constant,} \qquad TV(v) \le TV(u), \qquad \|u - v\|_{L^\infty(\mathbb{R})} \le \varepsilon.$$

Proof Consider the function

$$U : \mathbb{R} \longrightarrow \mathbb{R}, \qquad U(x) = TV(u\chi_{(-\infty,x]} + u(x^-)\chi_{(x,\infty)}). \qquad (5.3)$$

We have that

$$U \text{ is right continuous and nondecreasing,}$$
$$U(-\infty) = 0, \qquad U(\infty) = TV(u),$$
$$-\infty \le x \le y \le \infty \Longrightarrow |u(x) - u(y)| \le U(y) - U(x).$$

Define

$$x_0 = -\infty,$$
$$x_j = \min\{x \in \mathbb{R}; U(x) \ge j\varepsilon\}, \quad j = 1, \dots, N - 1,$$
$$x_N = \infty,$$

where

$$N = \left[\frac{TV(u)}{\varepsilon}\right].$$

Since U is nondecreasing

$$x_0 < \dots < x_N.$$

Consider the following piecewise constant function

$$v = u(-\infty)\chi_{(-\infty,x_1)} + \sum_{j=1}^{N-1} u(x_j)\chi_{[x_j,x_{j+1})}.$$

From the definition of v we gain

$$TV(v) = \sum_{j=0}^{N-2} |u(x_j) - u(x_{j+1})| \le TV(u).$$

We have to evaluate the L^∞ distance the between u and v. Let $x \in \mathbb{R}$. We distinguish two cases. If

$$x < x_1,$$

we have

$$v(x) = u(-\infty)$$

and from the definition of x_1

$$U(x) < \varepsilon.$$

Therefore

$$|u(x) - v(x)| = |u(x) - u(-\infty)| \le U(x) - U(-\infty) = U(x) < \varepsilon.$$

If

$$x_j \le x < x_{j+1}, \quad \text{for some } j \in \{1, .., N-1\},$$

we have

$$v(x) = u(x_j), \qquad U(x) < (j+1)\varepsilon, \qquad U(x_j) \ge j\varepsilon,$$
$$|u(x) - v(x)| = |u(x) - u(x_j)| \le U(x) - U(x_j) < \varepsilon.$$

$\square$

Theorem 5.4 *Let $u : \mathbb{R} \to \mathbb{R}$ and $\varepsilon > 0$. If*

$$u \text{ is right continuous}, \qquad u \in BV(\mathbb{R}),$$

$$u - u(-\infty) \in L^1(-\infty, 0), \qquad u - u(\infty) \in L^1(0, \infty),$$

then there exists a function $v : \mathbb{R} \to \mathbb{R}$ such that

$$v \text{ is piecewise constant,} \qquad TV(v) \leq TV(u), \qquad \|u - v\|_{L^1(\mathbb{R})} \leq \varepsilon.$$

Proof Since

$$u - u(-\infty) \in L^1(-\infty, 0), \qquad u - u(\infty) \in L^1(0, \infty)$$

there exists $\rho > 0$ such that

$$\int_{-\infty}^{-\rho} |u(x) - u(-\infty)| dx + \int_{\rho}^{\infty} |u(x) - u(\infty)| dx < \frac{\varepsilon}{2}.$$

Theorem 5.3 guarantees the existence of a piecewise constant function $\tilde{v}$ defined as follows

$$\tilde{v} = u(-\infty) \chi_{(-\infty, x_1)} + \sum_{j=1}^{N-1} u(x_j) \chi_{[x_j, x_{j+1})},$$

for some $-\infty < x_1 < \ldots < x_{N-1} < x_N = \infty$ and such that

$$TV(\tilde{v}) \leq TV(u), \qquad \|u - \tilde{v}\|_{L^\infty(\mathbb{R})} < \frac{\varepsilon}{4\rho}.$$

Define

$$v : \mathbb{R} \longrightarrow \mathbb{R}, \qquad v(x) = \begin{cases} u(\infty), & \text{if } x \geq \rho, \\ \tilde{v}(x), & \text{if } -\rho \leq x < \rho, \\ u(-\infty), & \text{if } x < -\rho. \end{cases}$$

Since

$$\text{Im}(v) \subset \overline{\text{Im}(u)},$$

we have

$$TV(v) \leq TV(u).$$

Moreover,

$$\|u - v\|_{L^1(\mathbb{R})} = \int_{-\infty}^{-\rho} |u(x) - v(x)| dx$$

$$+ \int_{-\rho}^{\rho} |u(x) - v(x)| dx + \int_{\rho}^{\infty} |u(x) - v(x)| dx$$

$$= \int_{-\infty}^{-\rho} |u(x) - u(-\infty)| dx$$

$$+ \int_{-\rho}^{\rho} |u(x) - \tilde{v}(x)| dx + \int_{\rho}^{\infty} |u(x) - u(\infty))| dx$$

$$\leq \int_{-\infty}^{-\rho} |u(x) - u(-\infty)| dx$$

$$+ 2\rho \, \|u - \tilde{v}\|_{L^{\infty}(\mathbb{R})} + \int_{\rho}^{\infty} |u(x) - u(\infty)| dx < \varepsilon.$$

$\square$

Theorem 5.5 *Let $u : \mathbb{R} \to \mathbb{R}$. If*

$$u \text{ is right continuous}, \qquad u \in BV(\mathbb{R}),$$

then

$$\frac{1}{\varepsilon} \int_{\mathbb{R}} |u(x + \varepsilon) - u(x)| dx \leq TV(u), \qquad \varepsilon > 0.$$

Proof Let $\varepsilon > 0$ and U be the function defined in (5.3). We have

$$\int_{\mathbb{R}} |u(x + \varepsilon) - u(x)| dx \leq \int_{\mathbb{R}} (U(x + \varepsilon) - U(x)) dx = \int_{\mathbb{R}} \left(\int_{U(x)}^{U(x+\varepsilon)} dy \right) dx$$

$$= \operatorname{meas} \left\{ (x, y) \in \mathbb{R}^2; U(x) \leq y \leq U(x + \varepsilon) \right\}$$

$$= \int_{U(-\infty)}^{U(\infty)} \operatorname{meas} \{ x \in \mathbb{R}; U(x) \leq y \leq U(x + \varepsilon) \} \, dy$$

$$= \int_{U(-\infty)}^{U(\infty)} \operatorname{meas} \left\{ x \in \mathbb{R}; x \leq U^{-1}(y) \leq x + \varepsilon \right\} dy$$

$$= \int_{U(-\infty)}^{U(\infty)} \operatorname{meas} \left\{ x \in \mathbb{R}; U^{-1}(y) - \varepsilon \leq x \leq U^{-1}(y) \right\} dy$$

$$= \int_{U(-\infty)}^{U(\infty)} \operatorname{meas} \left[U^{-1}(y) - \varepsilon, U^{-1}(y) \right] dy$$

$$= \int_{0}^{TV(u)} \varepsilon \, dy = \varepsilon TV(u).$$

$\square$

5.2 Compactness

This section is dedicated to some compactness properties of sequences uniformly bounded in BV.

Theorem 5.6 (Helly) *Let $\{u_n\}_{n\in\mathbb{N}} \subset BV(\mathbb{R})$ and $M, C > 0$. If*

$$TV(u_n) \leq C, \qquad \|u_n\|_{L^\infty(\mathbb{R})} \leq M, \qquad n \in \mathbb{N},$$

then there exist a subsequence $\{u_{n_k}\}_{k\in\mathbb{N}}$ and a function $u \in BV(\mathbb{R})$ such that

(i) $TV(u) \leq C$;
(ii) $\|u\|_{L^\infty(\mathbb{R})} \leq M$;
(iii) for every $x \in \mathbb{R}$, $u_{n_k}(x) \to u(x)$;
(iv) $u_{n_k} \to u$ strongly in $L^p_{loc}(\mathbb{R})$ for every $1 \leq p < \infty$.

Proof For every $n \in \mathbb{N}$, consider the function

$$U_n : \mathbb{R} \longrightarrow \mathbb{R}, \qquad U_n(x) = TV(u_n \chi_{(-\infty,x]} + u_n(x^-)\chi_{(x,\infty)}).$$

We have

$$U_n \text{ is nondecreasing,}$$

$$U_n(-\infty) = 0, \qquad U_n(\infty) = TV(u_n),$$

$$-\infty \leq x \leq y \leq \infty \Longrightarrow |u_n(x) - u_n(y)| \leq U_n(y) - U_n(x).$$

The Cantor diagonal method gives a subsequence $\{U_{n_k}\}_{k\in\mathbb{N}}$ of $\{U_n\}_{n\in\mathbb{N}}$ and a function U such that

$$\lim_k U_{n_k}(x) = U(x), \qquad x \in \mathbb{Q}.$$

Clearly U is nondecreasing and

$$0 \leq U(x) \leq C, \qquad x \in \mathbb{Q}.$$

Consider the sets

$$J_\nu = \left\{ x \in \mathbb{R}; \ \lim_{\substack{y\to x^+ \\ y\in\mathbb{Q}}} U(y) - \lim_{\substack{y\to x^- \\ y\in\mathbb{Q}}} U(y) \geq \frac{1}{\nu} \right\}, \qquad \nu \in \mathbb{N} \setminus \{0\},$$

$$J = \left\{ x \in \mathbb{R}; \ \lim_{\substack{y\to x^+ \\ y\in\mathbb{Q}}} U(y) \neq \lim_{\substack{y\to x^- \\ y\in\mathbb{Q}}} U(y) \right\}.$$

We have

$$J = \bigcup_{\nu=1}^{\infty} J_\nu,$$

and

$$\#(J_\nu) \leq C\nu.$$

Indeed

$$C \geq U(\infty) - U(-\infty) = TV(U)$$

$$\geq \sum_{x \in J_\nu} \left(\lim_{\substack{y \to x^+ \\ y \in \mathbb{Q}}} U(y) - \lim_{\substack{y \to x^- \\ y \in \mathbb{Q}}} U(y) \right) \geq \frac{\#(J_\nu)}{\nu}.$$

Using one more time the Cantor diagonal method, we gain a subsubsequence $\{u_{n_{k_\mu}}\}_{\mu \in \mathbb{N}}$ and a function u such that

$$\lim_\mu u_{n_{k_\mu}}(x) = u(x), \qquad x \in \mathbb{Q} \cup J.$$

We claim that $\{u_{n_{k_\mu}}\}_{\mu \in \mathbb{N}}$ converges to u in every point of $\mathbb{R}$. Indeed, if $x \in \mathbb{R} \setminus (\mathbb{Q} \cup J)$, we have

$$x \notin J_\nu, \qquad \nu \in \mathbb{N} \setminus \{0\}.$$

Let $\nu \in \mathbb{N} \setminus \{0\}$, there exist $p, q \in \mathbb{Q}$ such that

$$p < x < q, \qquad U(q) - U(p) < \frac{1}{\nu}.$$

Therefore,

$$|u(p) - u(q)| = \lim_\mu \left| u_{n_{k_\mu}}(p) - u_{n_{k_\mu}}(q) \right|$$

$$\leq \lim_\mu \left(U_{n_{k_\mu}}(q) - U_{n_{k_\mu}}(p) \right)$$

$$= U(q) - U(p) < \frac{1}{\nu},$$

and then $\{u_{n_{k_\mu}}\}_{\mu \in \mathbb{N}}$ converges to u in x.

Clearly we have

$$\|u\|_{L^\infty(\mathbb{R})} \leq M$$

and the Dominated Convergence Theorem gives

$$u_{n_{k_\mu}} \longrightarrow u \qquad \text{in } L^p_{loc}(\mathbb{R}) \text{ for every } 1 \le p < \infty.$$

It remains to prove that

$$u \in BV(\mathbb{R}), \qquad TV(u) \le M.$$

Let $x_0, \ldots, x_N \in \mathbb{R}$ be such that $x_0 < \ldots < x_N$. We have

$$\sum_{j=1}^{N} |u(x_j) - u(x_{j-1})| = \lim_\mu \sum_{j=1}^{N} \left| u_{n_{k_\mu}}(x_j) - u_{n_{k_\mu}}(x_{j-1}) \right|$$

$$\le \limsup_\mu TV(u_{n_{k_\mu}}) \le C.$$

$\square$

In light of Example 5.4 and Sobolev embedding $W^{1,1}(\mathbb{R}) \subset L^\infty(\mathbb{R})$, the following corollary is a immediate consequence of the Helly Theorem.

Corollary 5.1 *Let $\{u_n\}_{n \in \mathbb{N}} \subset W^{1,1}(\mathbb{R})$. If $\{u_n\}_{n \in \mathbb{N}}$ is uniformly bounded in $W^{1,1}(\mathbb{R})$ then there exist a subsequence $\{u_{n_k}\}_{k \in \mathbb{N}}$ and a function $u \in BV(\mathbb{R})$ such that* [1]

$$u_{n_k} \longrightarrow u \qquad \text{pointwise in } \mathbb{R} \text{ and strongly in } L^p_{loc}(\mathbb{R}) \text{ for every } 1 \le p < \infty.$$

We conclude the section with the following two dimensional version of the Helly Theorem.

Theorem 5.7 *Let $\{u_n\}_{n \in \mathbb{N}}$ be a sequence of functions defined on $[0, \infty) \times \mathbb{R}$ with value in $\mathbb{R}$ and $L, M, C > 0$ three constants. If*

$$\|u_n\|_{L^\infty((0,\infty) \times \mathbb{R})} \le M, \qquad n \in \mathbb{N},$$

$$TV(u_n(t, \cdot)) \le C, \qquad n \in \mathbb{N}, \, t \ge 0,$$

$$\|u_n(t, \cdot) - u_n(s, \cdot)\|_{L^1(\mathbb{R})} \le L|t - s|, \qquad n \in \mathbb{N}, \, t, s \ge 0,$$

then there exist a subsequence $\{u_{n_k}\}_{k \in \mathbb{N}}$ and a function $u : [0, \infty) \times \mathbb{R} \to \mathbb{R}$ such that

(i) $TV(u(t, \cdot)) \le C, \, t \ge 0$;
(ii) $\|u\|_{L^\infty((0,\infty) \times \mathbb{R})} \le M$;

---

[1] The Sobolev space $W^{1,1}(\mathbb{R})$ is not reflexive. As a consequence, in general, $u \notin W^{1,1}(\mathbb{R})$. Even if all the u_n are absolutely continuous in the limit u may be discontinuous.

(iii) $\|u(t, \cdot) - u(s, \cdot)\|_{L^1(\mathbb{R})} \leq L|t - s|$, $t, s \geq 0$;
(iv) for almost every $(t, x) \in (0, \infty) \times \mathbb{R}$, $u_{n_k}(t, x) \to u(t, x)$;
(v) $u_{n_k} \to u$ strongly in $L^p_{loc}((0, \infty) \times \mathbb{R})$ for every $1 \leq p < \infty$.

Proof Theorem 5.6 guarantees the existence of a subsequence $\{u_{n_k}\}_{k \in \mathbb{N}}$ and a function $u : ([0, \infty) \cap \mathbb{Q}) \times \mathbb{R} \to \mathbb{R}$ such that

$$TV(u(t, \cdot)) \leq C, \quad t \in [0, \infty) \cap \mathbb{Q},$$

$$\|u\|_{L^\infty(((0,\infty) \cap \mathbb{Q}) \times \mathbb{R})} \leq M,$$

$$\|u(t, \cdot) - u(s, \cdot)\|_{L^1(\mathbb{R})} \leq L|t - s|, \quad t, s \in [0, \infty) \cap \mathbb{Q},$$

$$u_{n_k}(t, \cdot) \longrightarrow u(t, \cdot) \quad \text{a.e. in } \mathbb{R} \text{ and every } t \in ([0, \infty) \cap \mathbb{Q}), \tag{5.4}$$

$$u_{n_k}(t, \cdot) \longrightarrow u(t, \cdot) \quad \text{in } L^p_{loc}(\mathbb{R}) \text{ for every } 1 \leq p < \infty \text{ and } t \in [0, \infty) \cap \mathbb{Q}.$$

The Lipschitz continuity in the time variable allows us to extend u to $[0, \infty) \times \mathbb{R}$ and gives (i), (ii), and (iii). (iv) follows from (5.4). Finally the Dominated Convergence Theorem and (iv) give (v). □

Corollary 5.2 Let $\{u_n\}_{n \in \mathbb{N}} \subset W^{1,1}([0, \infty) \times \mathbb{R})$ and $C > 0$. If

$$\|u_n(t, \cdot)\|_{L^1(\mathbb{R})} \leq C, \quad \|\partial_x u_n(t, \cdot)\|_{L^1(\mathbb{R})} \leq C, \quad \|\partial_t u_n(t, \cdot)\|_{L^1(\mathbb{R})} \leq C,$$

for every $t \geq 0$, then there exist a subsequence $\{u_{n_k}\}_{k \in \mathbb{N}}$ and a function $u : [0, \infty) \times \mathbb{R} \to \mathbb{R}$ such that

(i) $TV(u(t, \cdot)) \leq C$, $t \geq 0$;
(ii) $\|u(t, \cdot)\|_{L^1(\mathbb{R})} \leq C$, $t \geq 0$;
(iii) $\|u(t, \cdot) - u(s, \cdot)\|_{L^1(\mathbb{R})} \leq C|t - s|$, $t, s \geq 0$;
(iv) for almost every $(t, x) \in (0, \infty) \times \mathbb{R}$, $u_{n_k}(t, x) \to u(t, x)$;
(v) $u_{n_k} \to u$ strongly in $L^p_{loc}((0, \infty) \times \mathbb{R})$ for every $1 \leq p < \infty$.

Proof We have only to observe that, for every $t, s \geq 0$,

$$\|u_n(t, \cdot) - u_n(s, \cdot)\|_{L^1(\mathbb{R})} = \int_{\mathbb{R}} |u_n(t, x) - u_n(s, x)| dx$$

$$\leq \left| \int_s^t \int_{\mathbb{R}} |\partial_t u_n(\tau, x)| d\tau dx \right|$$

$$= \left| \int_s^t \|\partial_t u_n(\tau, \cdot)\|_{L^1(\mathbb{R})} d\tau \right| \leq C|t - s|$$

and then apply Theorem 5.7. □

5.3 Exercises

Exercise 5.1 Let μ be a finite Radon measure on $\mathbb{R}$ (see Definition 8.1). Prove that the function

$$f : \mathbb{R} \longrightarrow \mathbb{R}, \qquad f(x) = \mu((-\infty, x]),$$

has bounded variation.

Exercise 5.2 Prove that every real valued function with bounded variation is the difference of two nondecreasing functions.

Exercise 5.3 Give the details of the statement of Example 5.3.

Exercise 5.4 Give the details of the statement of Example 5.4.

Exercise 5.5 Let $\gamma : \mathbb{R} \to \mathbb{R}^n$ be a curve. Prove that γ is rectifiable if and only if all its components $\gamma_1, \ldots, \gamma_n$ have bounded variation.

Exercise 5.6 Prove that the function

$$f : \mathbb{R} \longrightarrow \mathbb{R}, \qquad f(x) = \begin{cases} \arctan(x) \sin(1/x), & \text{if } x \neq 0, \\ 0, & \text{if } x = 0, \end{cases}$$

is continuous, bounded but has unbounded variation.

Exercise 5.7 Prove that the product of two functions with bounded variation has bounded variation.

Exercise 5.8 Find two functions u, $v : \mathbb{R} \to \mathbb{R}$ such that

$$u, v \notin BV(\mathbb{R}), \qquad u^2 \in BV(\mathbb{R}), \qquad v^3 \in BV(\mathbb{R}).$$

Exercise 5.9 Let $\{u_n\}_{n \in \mathbb{N}}$ be a sequence of functions from $\mathbb{R}$ to $\mathbb{R}$ and $M, C > 0$. Prove that, if

$$TV(u_n^3) \leq C, \qquad \|u_n\|_{L^\infty(\mathbb{R})} \leq M, \qquad n \in \mathbb{N},$$

then there exist a subsequence $\{u_{n_k}\}_{k \in \mathbb{N}}$ and a function $u : \mathbb{R} \to \mathbb{R}$ such that

(i) $TV(u^3) \leq C$;
(ii) $\|u\|_{L^\infty(\mathbb{R})} \leq M$;
(iii) for every $x \in \mathbb{R}$, $u_{n_k}(x) \to u(x)$;
(iv) $u_{n_k} \to u$ strongly in $L^p_{loc}(\mathbb{R})$ for every $1 \leq p < \infty$.

Exercise 5.10 Find a sequence of functions $\{u_n\}_{n\in\mathbb{N}}$ from $\mathbb{R}$ to $\mathbb{R}$ such that

$$TV(u_n^2) \leq C, \qquad \|u_n\|_{L^\infty(\mathbb{R})} \leq M, \qquad n \in \mathbb{N},$$

for some constants $M, C > 0$ and no pointwise converging subsequences.

References

1. Ambrosio, L., Fusco, N., Pallara, D.: Functions of Bounded Variation and Free Discontinuity Problems. Clarendon Press, Oxford (2000)
2. Rudin, W.: Real and Complex Analysis, 3rd edn. McGraw-Hill Book Co., New York (1987)

Chapter 6
Wave Front Tracking

Abstract In this chapter we present the wave front-tracking approach to the existence issue. It consists in the approximation of conservations laws with equations of the same type but with piecewise affine fluxes and piecewise constant initial data. It has been extremely successful for instance in the analysis of systems and control problems.

Keywords Front-tracking · Existence

In Chap. 3 we proved the uniqueness and stability of entropy solutions of the Cauchy problems for conservation laws. Moreover, in Chap. 4, we started the analysis of the existence issue considering the "simple" case of the Riemann problems, that are Cauchy problems with Heaviside type initial conditions. In this chapter we prove the existence of solutions under the assumption that the initial condition has bounded variation. The algorithm that we develop here is called *front-tracking* and it has been introduced in [2]. It is based on the construction of piecewise constant approximate solutions. This algorithm has been used for the definition of numerical schemes [3] and it has been generalized to the case of systems [1].

6.1 Piecewise Constant Approximations

Let $n \in \mathbb{N}$ and f_n be the piecewise affine function coinciding with f in the points $2^{-n}i$, $i \in \mathbb{Z}$, namely

$$f_n(\xi) = f(2^{-n}i) + (\xi - 2^{-n}i)\frac{f(2^{-n}(i+1)) - f(2^{-n}i)}{2^{-n}},$$

$$2^{-n}i \leq \xi \leq 2^{-n}(i+1). \tag{6.1}$$

In addition, we consider a piecewise constant initial datum $u_{0,n}$ that takes values in $2^{-n}\mathbb{Z}$. In this section, we prove that the initial value problem

$$\begin{cases} \partial_t u + \partial_x f_n(u) = 0, & t > 0,\ x \in \mathbb{R}, \\ u(0, x) = u_{0,n}(x), & x \in \mathbb{R}, \end{cases} \tag{6.2}$$

admits an entropy solution $u = u(t, x)$ such that

(i) $u(t, \cdot)$ is piecewise constant for every $t \geq 0$;
(ii) u takes values in $2^{-n}\mathbb{Z}$;
(iii) $TV(u(t, \cdot)) \leq TV(u_{0,n})$ for every $t \geq 0$;
(iv) $\|u(t, \cdot)\|_{L^\infty(\mathbb{R})} \leq \|u_{0,n}\|_{L^\infty(\mathbb{R})}$ for every $t \geq 0$.

Let

$$x_1 < \ldots . < x_N$$

be the jump points of $u_{0,n}$. In correspondence of every x_i, we solve the Riemann problem with left and right states $u_{0,n}(x_i^-)$ and $u_{0,n}(x_i^+)$. In this way we obtain a solution defined for small enough $t > 0$. This solution can be defined up to the first time $t_1 > 0$ when two or more discontinuities generated at time $t = 0$ cross. Since u takes values in $2^{-n}\mathbb{Z}$ and is piecewise constant, we can use $u(t_1^-, \cdot)$ as a new initial condition. Therefore we solve the Riemann problems at time $t = t_1$ and get a solution defined until a second interaction occurs at time t_2 (see Fig. 6.1). We have to show that this algorithm allows us to construct a solution u defined for every time $t > 0$. It suffices to prove that there are finitely many interaction in every interval $[0, T]$. Indeed, let

$$\xi_1(t) < \ldots . < \xi_m(t)$$

Fig. 6.1 Wave fronts

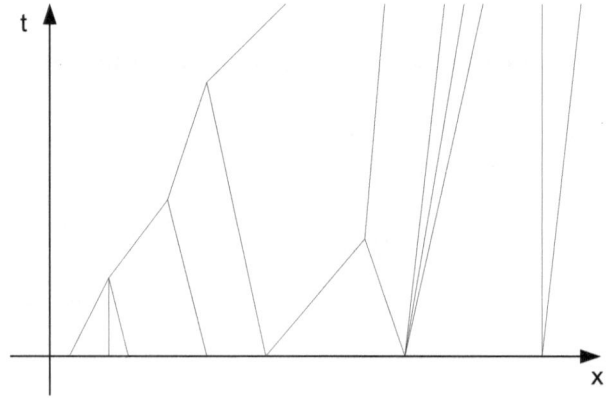

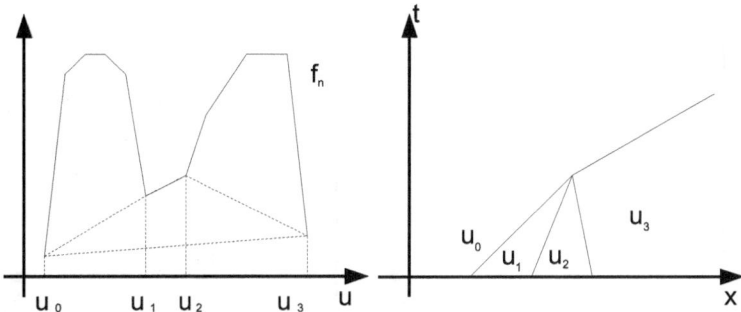

Fig. 6.2 Fronts interaction

be the position of m discontinuities that interact at time τ. Let $u_0, \ldots, u_m$ be the values that u takes before the interaction at times $t < \tau$, namely

$$u_0 = u(t, \xi_1(t)^-), \quad u_i = u(t, \xi_i(t)^+), \qquad i = 1, \ldots, m.$$

We distinguish two cases. Assume that all the jumps $u_i - u_{i-1}$ have the same sign. We claim that after the interaction we get only one front connecting u_0 and u_m. It is not restrictive to consider only the case in which all the jumps have positive sign (see Fig. 6.2), i.e.,

$$u_0 < u_1 < \ldots < u_m.$$

Since u is an entropy solution, we have

$$\xi_i' = \frac{f_n(u_i) - f(u_{i-1})}{u_i - u_{i-1}},$$

$$f_n(s) \geq (s - u_i)\frac{f_n(u_i) - f(u_{i-1})}{u_i - u_{i-1}} + f_n(u_i), \quad u_{i-1} \leq s \leq u_i,$$

for every $i \in \{1, \ldots, m\}$. Since all the fronts meet at the same point, we must have

$$\xi_1' > \xi_2' > \ldots > \xi_m',$$

therefore

$$f_n(s) \geq (s - u_0)\frac{f_n(u_m) - f(u_0)}{u_m - u_0} + f_n(u_0), \quad u_0 \leq s \leq u_m.$$

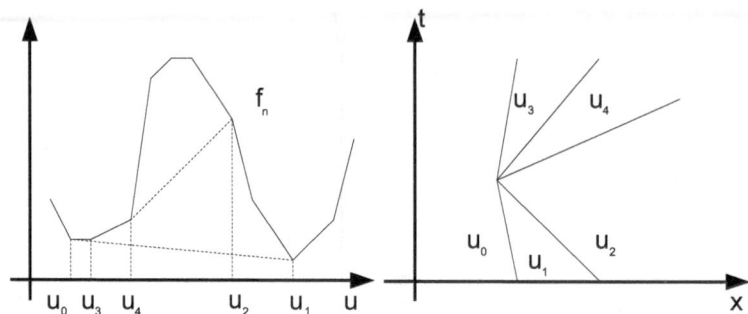

Fig. 6.3 Fronts interaction

As a consequence the shock (u_0, u_m) with speed

$$\frac{f_n(u_m) - f(u_0)}{u_m - u_0}$$

satisfies the entropy conditions. Due to the uniqueness of the entropy solutions, after time τ we only have the shock (u_0, u_m). Finally, observe that the number of fronts decreased and the total variation and L^∞ norm did no change.

Let us consider now the case in which at least two jumps of the interacting fronts have different sign (see Fig. 6.3). In this case the number of fronts may increase after the interaction. Anyway since the total amplitude of the new fronts is $|u_0 - u_m|$, the total variation of the solution is decreasing at least of $2 \cdot 2^{-n}$. In Fig. 6.3, we have an example of two interacting fonts ($m = 2$) with jumps having different sign. The Riemann problem generated by the interaction generates three fronts connecting the states $u_0 < u_3 < u_4 < u_2$.

Since the total variation of $u(t, \cdot)$ is bounded at $t = 0$, it is not increasing with the interactions and every jump is proportional to 2^{-n}, the number of interactions must stay bounded in every time interval $[0, T]$, $T > 0$.

6.2   Global Existence of BV Solutions

Here we prove the existence of entropy solutions for initial value problems for conservation laws with initial datum in BV.

Theorem 6.1 *Let $u_0 \in L^1(\mathbb{R}) \cap BV(\mathbb{R})$ and $f \in C^2(\mathbb{R})$. The Cauchy problem*

$$\begin{cases} \partial_t u + \partial_x f(u) = 0, & t > 0, \ x \in \mathbb{R}, \\ u(0, x) = u_0(x), & x \in \mathbb{R}, \end{cases} \tag{6.3}$$

admits a unique entropy solution such that

$$TV(u(t, \cdot)) \leq TV(u_0), \qquad \|u(t, \cdot)\|_{L^\infty(\mathbb{R})} \leq \|u_0\|_{L^\infty(\mathbb{R})},$$

for every $t \geq 0$.

Proof Since $u_0 \in L^1(\mathbb{R}) \cap BV(\mathbb{R})$, there exists a sequence of piecewise constant functions $\{u_{0,n}\}_{n \in \mathbb{N}}$ such that

  (i) $u_{0,n}$ tales values in $2^{-n}\mathbb{Z}$ for every $n \in \mathbb{N}$;
  (ii) $u_{0,n} \to u_0$ in $L^1(\mathbb{R})$ as $n \to \infty$;
  (iii) $TV(u_{0,n}) \leq TV(u_0)$ for every $n \in \mathbb{N}$;
  (iv) $\|u_{0,n}\|_{L^\infty(\mathbb{R})} \leq \|u_0\|_{L^\infty(\mathbb{R})}$ for every $n \in \mathbb{N}$.

In correspondence of every n we approximate f with a piecewise affine function f_n (see (6.1)). Let $u_n = u_n(t, x)$ be the entropy solution of the Cauchy problem

$$\begin{cases} \partial_t u_n + \partial_x f_n(u_n) = 0, & t > 0, \ x \in \mathbb{R}, \\ u_n(0, x) = u_{0,n}(x), & x \in \mathbb{R}. \end{cases}$$

We have that

$$u_n(t, x) \in 2^{-n}\mathbb{Z}, \qquad t \geq 0, \ x \in \mathbb{R}, \ n \in \mathbb{N},$$

$$TV(u_n(t, \cdot)) \leq TV(u_{0,n}) \leq TV(u_0), \qquad t \geq 0, \ n \in \mathbb{N}, \tag{6.4}$$

$$\|u_n(t, \cdot)\|_{L^\infty(\mathbb{R})} \leq \|u_{0,n}\|_{L^\infty(\mathbb{R})} \leq \|u_0\|_{L^\infty(\mathbb{R})}, \qquad t \geq 0, \ n \in \mathbb{N}.$$

We claim that

$$\|u_n(t, \cdot) - u_n(s, \cdot)\|_{L^1(\mathbb{R})} \leq L|t - s|TV(u_0), \qquad t, s \geq 0, \ n \in \mathbb{N}, \tag{6.5}$$

where

$$L = \sup |f'(u_0)|.$$

It is not restrictive to consider only the case $t > s \geq 0$. We have

$$\|u_n(t, \cdot) - u_n(s, \cdot)\|_{L^1(\mathbb{R})} = \int_{\mathbb{R}} |u_n(t, x) - u_n(s, x)| dx$$

$$\leq \int_{\mathbb{R}} \int_s^t |\partial_t u_n(\tau, x)| d\tau dx$$

$$= \int_{\mathbb{R}} \int_s^t |\partial_x f(u_n(\tau, x))| d\tau dx$$

$$\underset{\text{see (6.4)}}{\leq} L \int_{\mathbb{R}} \int_s^t |\partial_x u_n(\tau, x)| d\tau dx$$

$$= L \int_s^t TV(u_n(\tau, \cdot)) d\tau$$

$$\leq L|t - s| TV(u_0).$$

Therefore, Theorem 5.7 guarantees the existence of a subsequence $\{u_{n_k}\}_{k \in \mathbb{N}}$ and a function

$$u \in L^1_{loc}((0, \infty) \times \mathbb{R}) \cap L^\infty(0, \infty; BV(\mathbb{R})),$$

such that

$$u_{n_k} \longrightarrow u \quad \text{a.e. and in } L^1_{loc}((0, \infty) \times \mathbb{R}) \text{ as } k \to \infty,$$

$$TV(u(t, \cdot)) \leq TV(u_0), \qquad t \geq 0,$$

$$\|u(t, \cdot)\|_{L^\infty(\mathbb{R})} \leq \|u_0\|_{L^\infty(\mathbb{R})}, \qquad t \geq 0.$$

Since

$$f_n \longrightarrow f \quad \text{uniformly in } [-\|u_0\|_{L^\infty(\mathbb{R})}, \|u_0\|_{L^\infty(\mathbb{R})}],$$

thanks to the Dominated Convergence Theorem u is the entropy solution of (6.2).

□

We conclude this section with the following theorem on the semigroup of entropy solutions associated tot he Cauchy problem (6.3)

Theorem 6.2 *There exists a continuous semigroup of operators $S : [0, \infty) \times L^1(\mathbb{R}) \to L^1(\mathbb{R})$ such that*

- *for every $t, s \geq 0$ and $u_0 \in L^1(\mathbb{R})$, $S_0(u_0) = u_0$, $S_t(S_s(u_0)) = S_{t+s}(u_0)$;*
- *for every $t \geq 0$ and $u_0, v_0 \in L^1(\mathbb{R})$, $\|S_t(u_0) - S_t(v_0)\|_{L^1(\mathbb{R})} \leq \|u_0 - v_0\|_{L^1(\mathbb{R})}$;*
- *for every $u_0 \in L^1(\mathbb{R}) \cap L^\infty(\mathbb{R})$ then $u(t, x) = S_t(u_0)(x)$ is the unique entropy solution of (6.3) in correspondence of the initial condition u_0;*
- *for every $t \geq 0$ and $u_0 \in L^1(\mathbb{R}) \cap L^\infty(\mathbb{R})$, $\sup_{t \geq 0} \|S_t(u_0)\|_{L^\infty(\mathbb{R})} \leq \|u_0\|_{L^\infty(\mathbb{R})}$;*
- *for every $t \geq 0$ and $u_0 \in L^1(\mathbb{R}) \cap BV(\mathbb{R})$, $\sup_{t \geq 0} TV(S_t(u_0)) \leq TV(u_0)$.*

Proof For every $u_0 \in L^1(\mathbb{R}) \cap BV(\mathbb{R})$ let $S_t(u_0) = u(t, \cdot)$ where u is the unique entropy solution of (6.3) in correspondence of the initial condition u_0 (see Theorems 3.9 and 6.1). We have that

$$S_0(u_0) = u_0, \tag{6.6}$$

$$S_t(S_s(u_0)) = S_{t+s}(u_0), \tag{6.7}$$

$$\sup_{t\geq 0} \|S_t(u_0)\|_{L^\infty(\mathbb{R})} \leq \|u_0\|_{L^\infty(\mathbb{R})}, \tag{6.8}$$

$$\sup_{t\geq 0} TV(S_t(u_0)) \leq TV(u_0), \tag{6.9}$$

$$\|S_t(u_0) - S_t(v_0)\|_{L^1(\mathbb{R})} \leq \|u_0 - v_0\|_{L^1(\mathbb{R})}, \tag{6.10}$$

for every $u_0, v_0 \in L^1(\mathbb{R}) \cap BV(\mathbb{R})$ and $t, s \geq 0$. Given $u_0 \in L^1(\mathbb{R})$, consider a sequence $\{u_{0,n}\}_{n\in\mathbb{N}} \subset L^1(\mathbb{R}) \cap BV(\mathbb{R})$, such that

$$u_{0,n} \to u_0 \qquad \text{strongly in } L^1(\mathbb{R}) \tag{6.11}$$

and define

$$S(u_0) = \lim_n S(u_{0,n}).$$

Thanks to (6.10) and (6.11), the definition of $S(u_0)$ is independent on the choice of the sequence $\{u_{0,n}\}_{n\in\mathbb{N}}$. The claim follows from (6.6), (6.7), (6.8), (6.9), (6.10), (6.11), and the Dominated Convergence Theorem. $\qquad\qquad\square$

6.3 Exercises

Exercise 6.1 Find the wave front tracking approximant of the solution of the Riemann problem

$$
\begin{cases}
\partial_t u + \partial_x \left(\frac{u^2}{2}\right) = 0, & t > 0,\ x \in \mathbb{R}, \\
u(0, x) = \begin{cases} 1, & \text{if } x \geq 0, \\ 0, & \text{if } x < 0, \end{cases}
\end{cases}
$$

with values in $2^{-3}\mathbb{Z}$.

References

1. Bressan, A.:   Hyperbolic Systems of Conservation Laws – The One-Dimensional Cauchy Problem. Oxford Lecture Series in Mathematics and its Applications, vol. 20. Oxford University Press, Oxford (2000)
2. Dafermos, C.M.: Polygonal approximations of solutions of the initial value problem for a conservation law. J. Math. Anal. Appl. **38**, 33–41 (1972)
3. Holden, H., Risebro, N.H.:   Front Tracking for Hyperbolic Conservation Laws. Applied Mathematical Sciences, vol. 152, 2nd edn. Springer, Heidelberg (2015)

Chapter 7
Vanishing Viscosity

Abstract In this chapter we present the vanishing viscosity approach to the existence issue. It is based on a parabolic approximation of the hyperbolic equation. It is interesting for the reminiscence of the convergence of the Navier-Stokes equations towards the Euler ones in fluid-dynamics. Moreover, it gives precious hints on the convergence analysis of numerical schemes.

Keywords Vanishing viscosity · Error estimate · Kuznecov Lemma · Existence · Parabolic approximation

In this chapter we discuss the parabolic approximation

$$\begin{cases} \partial_t u_\varepsilon + \partial_x f(u_\varepsilon) = \varepsilon \partial_{xx}^2 u_\varepsilon, & t > 0,\ x \in \mathbb{R}, \\ u_\varepsilon(0, x) = u_{0,\varepsilon}(x), & x \in \mathbb{R}, \end{cases} \tag{7.1}$$

of the scalar hyperbolic conservation law

$$\begin{cases} \partial_t u + \partial_x f(u) = 0, & t > 0,\ x \in \mathbb{R}, \\ u(0, x) = u_0(x), & x \in \mathbb{R}. \end{cases} \tag{7.2}$$

The mean feature of such an approximation is in the regularity of the solutions. Indeed due to its parabolic structure (7.1) does not experience shocks.

On (7.2) we assume

$$u_0 \in L^1(\mathbb{R}) \cap BV(\mathbb{R}).$$

On the other hand, for every $\varepsilon > 0$, $u_{0,\varepsilon}$ is a smooth approximation to u_0 such that

$$
\begin{aligned}
&u_{0,\varepsilon} \in C^\infty(\mathbb{R}) \cap W^{2,1}(\mathbb{R}), \qquad \varepsilon > 0, \\
&u_{0,\varepsilon} \longrightarrow u_0, \qquad \text{in } L^p(\mathbb{R}), \ 1 \le p < \infty, \text{ as } \varepsilon \to 0, \\
&\left\| u_{0,\varepsilon} \right\|_{L^\infty(\mathbb{R})} \le \left\| u_0 \right\|_{L^\infty(\mathbb{R})}, \quad \left\| u_{0,\varepsilon} \right\|_{L^1(\mathbb{R})} \le \left\| u_0 \right\|_{L^1(\mathbb{R})}, \qquad \varepsilon > 0, \\
&\left\| \partial_x u_{0,\varepsilon} \right\|_{L^1(\mathbb{R})} \le TV(u_0), \quad \varepsilon \left\| \partial^2_{xx} u_{0,\varepsilon} \right\|_{L^1(\mathbb{R})} \le C, \qquad \varepsilon > 0,
\end{aligned}
\tag{7.3}
$$

for some constant $C > 0$ independent on ε. Under these assumptions (7.1) admits a unique solution u_ε such that

$$
u_\varepsilon \in C^\infty([0,\infty) \times \mathbb{R}) \cap W^{2,p}(0,\infty; W^{1,p}(\mathbb{R})), \qquad 1 \le p < \infty.
$$

The main result of this chapter is the following [3, 6].

Theorem 7.1 *If*

$$
u_0 \in L^1(\mathbb{R}) \cap BV(\mathbb{R}),
$$

then

$$
u_\varepsilon \longrightarrow u \qquad \text{in } L^p_{loc}((0,\infty) \times \mathbb{R}) \text{ and a.e. in } (0,\infty) \times \mathbb{R},
\tag{7.4}
$$

where u is the entropy weak solution of (7.2) and u_ε is the solution of (7.1). Moreover, the following estimate holds

$$
\left\| u_\varepsilon(t,\cdot) - u(t,\cdot) \right\|_{L^1(\mathbb{R})} \le c\sqrt{\varepsilon t}\, TV(u_0) + \left\| u_{0,\varepsilon} - u_0 \right\|_{L^1(\mathbb{R})},
\tag{7.5}
$$

for every $\varepsilon > 0$ and $t \ge 0$, where c is a positive constant independent on ε and t.

The convergence part of this result has been proved in [4] for scalar equations and in [2] for systems of conservation laws. The error estimate (7.5) has been proved in [5].

Let us conclude this introduction with the following observation. In our statement all the family $\{u_\varepsilon\}_{\varepsilon > 0}$ converges to u and not just a subsequences: this follows from the uniqueness of the entropy solutions of (7.2) and the following equivalence

$$
u_\varepsilon \longrightarrow u
$$

$$
\Updownarrow
\tag{7.6}
$$

$$
\forall \, \{u_{\varepsilon_k}\}_{k \in \mathbb{N}} \text{ subsequence } \exists \, \{u_{\varepsilon_{k_h}}\}_{h \in \mathbb{N}} \text{ subsequence s.t. } u_{\varepsilon_{k_h}} \longrightarrow u.
$$

7.1   A Priori Estimates, Compactness, and Convergence

This section is dedicated to the proof of (7.4). Let us start with a technical lemma that will play a key role in the following a priori estimates.

Lemma 7.1 ([1, Lemma 2]) *Let $v : \mathbb{R} \to \mathbb{R}$ be a function. If*

$$v \in C^1(\mathbb{R}), \qquad v' \in L^1(\mathbb{R}),$$

then

$$\lim_{\delta \to 0+} \int_{|v| < \delta} |v'| dx = 0.$$

Proof We write

$$v_\delta = |v'| \chi_{\{|v| < \delta\}}, \qquad \delta > 0,$$

and observe that

$$|v_\delta| \le |v'|, \qquad v_\delta \longrightarrow 0 \quad \text{q.o. in } \mathbb{R}.$$

Indeed, if $|\{v = 0\}| = 0$ we have $\chi_{\{|v| < \delta\}} \to 0$ otherwise $v' \to 0$ on $\{v = 0\}$. Therefore the claim follows from the Dominated Convergence Theorem. $\square$

Remark 7.1 Since the solutions of (7.1) are smooth, the previous lemma allows us to use the identity

$$\text{sign}(v)' = \delta_{\{v=0\}} v', \tag{7.7}$$

in our computations, where $\delta_{\{v=0\}}$ is the Dirac delta concentrated on the set $\{v = 0\}$. In particular, if $v \in C^2(\mathbb{R}) \cap L^\infty(\mathbb{R}) \cap W^{2,1}(\mathbb{R})$, we have

$$\int_{\mathbb{R}} f(v)'' \text{sign}(v') \, dx = 0, \qquad \int_{\mathbb{R}} v'' \text{sign}(v) \, dx \le 0, \tag{7.8}$$

that follow integrating by parts and using (7.7).

Let us give a rigorous proof of them. We have

$$\lim_{\alpha \to 0} \int_{\mathbb{R}} f(v)'' \eta'_\alpha(v') dx = \int_{\mathbb{R}} f(v)'' \text{sign}(v') \, dx,$$

$$\lim_{\alpha \to 0} \int_{\mathbb{R}} v'' \eta'_\alpha(v) dx = \int_{\mathbb{R}} v'' \text{sign}(v) \, dx, \tag{7.9}$$

where

$$\eta_\alpha(\xi) = \sqrt{\xi^2 + \alpha^2}, \qquad \alpha \in \mathbb{R}.$$

For every $\alpha \neq 0$

$$\eta_\alpha \in C^2(\mathbb{R}), \qquad \eta'_\alpha(\xi) = \frac{\xi}{\sqrt{\xi^2 + \alpha^2}}, \qquad \eta''_\alpha(\xi) = \frac{\alpha^2}{(\xi^2 + \alpha^2)^{3/2}} \geq 0.$$

We have

$$\left| \int_{\mathbb{R}} f(v)'' \eta'_\alpha(v') dx \right| = \left| \int_{\mathbb{R}} f'(v) v' \eta''_\alpha(v') v'' dx \right|$$

$$\leq L \int_{\mathbb{R}} \left| v' \eta''_\alpha(v') v'' \right| dx$$

$$= L \int_{\mathbb{R}} \left| \frac{\alpha^2 v' v''}{((v')^2 + \alpha^2)^{3/2}} \right| dx$$

$$= L \int_{\{|v'| < \sqrt{\alpha}\}} \left| \frac{\alpha^2 v' v''}{((v')^2 + \alpha^2)^{3/2}} \right| dx$$

$$+ L \int_{\{|v'| \geq \sqrt{\alpha}\}} \left| \frac{\alpha^2 v' v''}{((v')^2 + \alpha^2)^{3/2}} \right| dx$$

$$\leq \frac{3}{8} L \underbrace{\int_{\{|v'| < \sqrt{\alpha}\}} |v''| dx}_{\to 0 \text{ by Lemma 7.1}}$$

$$+ L \underbrace{\frac{\alpha}{(1 + \alpha)^{3/2}}}_{\to 0} \int_{\{|v'| \geq \sqrt{\alpha}\}} |v''| dx \longrightarrow 0,$$

$$\int_{\mathbb{R}} v'' \eta'_\alpha(v) dx = - \int_{\mathbb{R}} \eta''_\alpha(v)(v')^2 dx \leq 0,$$

where $L = \sup\limits_{|\xi| \leq \|v\|_{L^\infty(\mathbb{R})}} |f'(\xi)|$. Therefore, (7.8) follows from (7.9).

Let us continue with some a priori estimates on u_ε, independent on ε.

Lemma 7.2 (L^∞ Estimate) *We have that*

$$\|u_\varepsilon\|_{L^\infty((0,\infty) \times \mathbb{R})} \leq \|u_0\|_{L^\infty(\mathbb{R})}, \qquad \varepsilon > 0.$$

Proof Due to (7.3) the maps with constant values $\|u_0\|_{L^\infty(\mathbb{R})}$ and $-\|u_0\|_{L^\infty(\mathbb{R})}$ provide a super and a sub solution to (7.1), respectively. Therefore, the claim follows from the comparison principle for parabolic equations. □

Lemma 7.3 (L^1 Estimate) *The function*

$$t \geq 0 \longmapsto \|u_\varepsilon(t, \cdot)\|_{L^1(\mathbb{R})}$$

is nonincreasing. In particular,

$$\|u_\varepsilon(t, \cdot)\|_{L^1(\mathbb{R})} \leq \|u_0\|_{L^1(\mathbb{R})}, \qquad \varepsilon > 0, \quad t \geq 0.$$

Proof Due to the regularity of u_ε, we have

$$\frac{d}{dt} \int_{\mathbb{R}} |u_\varepsilon| dx = \int_{\mathbb{R}} \operatorname{sign}(u_\varepsilon) \, \partial_t u_\varepsilon dx$$

$$= \varepsilon \int_{\mathbb{R}} \operatorname{sign}(u_\varepsilon) \, \partial_{xx}^2 u_\varepsilon dx - \int_{\mathbb{R}} \operatorname{sign}(u_\varepsilon) \, f'(u_\varepsilon) \partial_x u_\varepsilon dx$$

$$= \underbrace{-\varepsilon \int_{\mathbb{R}} \delta_{\{u_\varepsilon=0\}} \, (\partial_x u_\varepsilon)^2 \, dx}_{\leq 0} - \underbrace{\int_{\mathbb{R}} \partial_x \left(\int_0^{u_\varepsilon(t,x)} \operatorname{sign}(s) \, f'(s) ds \right) dx}_{=0} \leq 0,$$

where $\delta_{\{u_\varepsilon=0\}}$ is the Dirac's delta concentrated on the set $\{u_\varepsilon = 0\}$. Finally, an integration on $(0, t)$ gives (see (7.3))

$$\|u_\varepsilon(t, \cdot)\|_{L^1(\mathbb{R})} \leq \|u_{0,\varepsilon}\|_{L^1(\mathbb{R})} \leq \|u_0\|_{L^1(\mathbb{R})}.$$

□

Lemma 7.4 (BV Estimate in x) *The function*

$$t \geq 0 \longmapsto \|\partial_x u_\varepsilon(t, \cdot)\|_{L^1(\mathbb{R})}$$

is nonincreasing. In particular,

$$\|\partial_x u_\varepsilon(t, \cdot)\|_{L^1(\mathbb{R})} \leq TV(u_0), \qquad \varepsilon > 0, \quad t \geq 0.$$

Proof Due to the regularity of u_ε, we have

$$\partial_{tx}^2 u_\varepsilon + \partial_x \left(f'(u_\varepsilon) \partial_x u_\varepsilon \right) = \varepsilon \partial_{xxx}^3 u_\varepsilon,$$

and then

$$\frac{d}{dt} \int_{\mathbb{R}} |\partial_x u_\varepsilon| dx = \int_{\mathbb{R}} \text{sign}\,(\partial_x u_\varepsilon)\, \partial_{tx}^2 u_\varepsilon dx$$

$$=\varepsilon \int_{\mathbb{R}} \text{sign}\,(\partial_x u_\varepsilon)\, \partial_{xxx}^3 u_\varepsilon dx - \int_{\mathbb{R}} \text{sign}\,(\partial_x u_\varepsilon)\, \partial_x \left(f'(u_\varepsilon)\partial_x u_\varepsilon \right) dx$$

$$= -\varepsilon \underbrace{\int_{\mathbb{R}} \delta_{\{\partial_x u_\varepsilon = 0\}} \left(\partial_{xx}^2 u_\varepsilon \right)^2 dx}_{\leq 0} + \underbrace{\int_{\mathbb{R}} \delta_{\{\partial_x u_\varepsilon = 0\}} \partial_{xx}^2 u_\varepsilon\, f'(u_\varepsilon) \partial_x u_\varepsilon dx}_{=0} \leq 0,$$

where $\delta_{\{\partial_x u_\varepsilon = 0\}}$ is the Dirac's delta concentrated on the set $\{\partial_x u_\varepsilon = 0\}$. Finally, an integration on $(0, t)$ gives (see (7.3))

$$\|\partial_x u_\varepsilon(t, \cdot)\|_{L^1(\mathbb{R})} \leq \|\partial_x u_{0,\varepsilon}\|_{L^1(\mathbb{R})} \leq TV(u_0).$$

$\square$

Lemma 7.5 (*BV* **Estimate in** t) *The function*

$$t \geq 0 \longmapsto \|\partial_t u_\varepsilon(t, \cdot)\|_{L^1(\mathbb{R})}$$

is nonincreasing. In particular,

$$\|\partial_t u_\varepsilon(t, \cdot)\|_{L^1(\mathbb{R})} \leq TV(u_0)L + C, \qquad \varepsilon > 0, \quad t \geq 0,$$

where C is the constant that appears in (7.3) and

$$L = \|f'\|_{L^\infty(-\|u_0\|_{L^\infty(\mathbb{R})}, \|u_0\|_{L^\infty(\mathbb{R})})}.$$

Proof Due to the regularity of u_ε, we have

$$\partial_{tt}^2 u_\varepsilon + \partial_x \left(f'(u_\varepsilon)\partial_t u_\varepsilon \right) = \varepsilon \partial_{txx}^3 u_\varepsilon,$$

and then

$$\frac{d}{dt} \int_{\mathbb{R}} |\partial_t u_\varepsilon| dx = \int_{\mathbb{R}} \text{sign}\,(\partial_t u_\varepsilon)\, \partial_{tt}^2 u_\varepsilon dx$$

$$=\varepsilon \int_{\mathbb{R}} \text{sign}\,(\partial_t u_\varepsilon)\, \partial_{txx}^3 u_\varepsilon dx - \int_{\mathbb{R}} \text{sign}\,(\partial_t u_\varepsilon)\, \partial_x \left(f'(u_\varepsilon)\partial_t u_\varepsilon \right) dx$$

$$= -\varepsilon \underbrace{\int_{\mathbb{R}} \delta_{\{\partial_t u_\varepsilon = 0\}} \left(\partial_{tx}^2 u_\varepsilon \right)^2 dx}_{\leq 0} + \underbrace{\int_{\mathbb{R}} \delta_{\{\partial_t u_\varepsilon = 0\}} \partial_{tx}^2 u_\varepsilon\, f'(u_\varepsilon) \partial_t u_\varepsilon dx}_{=0} \leq 0,$$

where $\delta_{\{\partial_t u_\varepsilon = 0\}}$ is the Dirac's delta concentrated on the set $\{\partial_t u_\varepsilon = 0\}$. Finally, an integration on $(0, t)$, (7.1), (7.3), and Lemma 7.2 give

$$\|\partial_t u_\varepsilon(t, \cdot)\|_{L^1(\mathbb{R})} \leq \|\partial_t u_\varepsilon(0, \cdot)\|_{L^1(\mathbb{R})}$$

$$= \left\|\varepsilon \partial_{xx}^2 u_{0,\varepsilon} - f'(u_{0,\varepsilon})\partial_x u_{0,\varepsilon}\right\|_{L^1(\mathbb{R})}$$

$$\leq \varepsilon \left\|\partial_{xx}^2 u_{0,\varepsilon}\right\|_{L^1(\mathbb{R})} + \left\|f'(u_{0,\varepsilon})\right\|_{L^\infty(\mathbb{R})}\left\|\partial_x u_{0,\varepsilon}\right\|_{L^1(\mathbb{R})}$$

$$\leq C + TV(u_0)L.$$

$\square$

Proof of Eq. (7.4) Let $\{u_{\varepsilon_k}\}_{k \in \mathbb{N}}$ be a subsequence of $\{u_\varepsilon\}_{\varepsilon > 0}$. Since $\{u_{\varepsilon_k}\}_{k \in \mathbb{N}}$ is bounded in $L^\infty((0, \infty) \times \mathbb{R}) \cap BV((0, T) \times \mathbb{R})$, $T > 0$, (see Lemmas 7.3, 7.4, and 7.5), there exists a function $u \in L^\infty((0, \infty) \times \mathbb{R}) \cap BV((0, T) \times \mathbb{R})$, $T > 0$, and a subsequence $\{u_{\varepsilon_{k_h}}\}_{h \in \mathbb{N}}$ such that

$$u_{\varepsilon_{k_h}} \longrightarrow u \qquad \text{in } L^p_{loc}((0, \infty) \times \mathbb{R}) \text{ and a.e. in}(0, \infty) \times \mathbb{R}.$$

We claim that u is the unique entropy solution of (7.2). Let $\eta \in C^2(\mathbb{R})$ be a convex entropy with flux q defined by $q' = \eta' f'$. Multiplying (7.1) by $\eta'(u_{\varepsilon_{k_h}})$ we get

$$\partial_t \eta(u_{\varepsilon_{k_h}}) + \partial_x q(u_{\varepsilon_{k_h}}) = \varepsilon_{k_h} \partial_{xx}^2 u_{\varepsilon_{k_h}} \eta'(u_{\varepsilon_{k_h}})$$

$$= \varepsilon_{k_h} \partial_{xx}^2 \eta(u_{\varepsilon_{k_h}}) \underbrace{-\varepsilon_{k_h} \eta''(u_{\varepsilon_{k_h}})(\partial_x u_{\varepsilon_{k_h}})^2}_{\leq 0}$$

$$\leq \varepsilon_{k_h} \partial_{xx}^2 \eta(u_{\varepsilon_{k_h}}).$$

For every nonnegative test function $\varphi \in C^\infty(\mathbb{R}^2)$ with compact support, we have that

$$\int_0^\infty \int_{\mathbb{R}} \left(\eta(u_{\varepsilon_{k_h}})\partial_t \varphi + q(u_{\varepsilon_{k_h}})\partial_x \varphi\right) dt dx + \int_{\mathbb{R}} \eta(u_{0,\varepsilon_{k_h}}(x))\varphi(0, x)dx$$

$$\geq -\varepsilon_{k_h} \int_0^\infty \int_{\mathbb{R}} \eta(u_{\varepsilon_{k_h}})\partial_{xx}^2 \varphi dt dx.$$

As $h \to \infty$, the Dominated Convergence Theorem gives

$$\int_0^\infty \int_{\mathbb{R}} (\eta(u)\partial_t \varphi + q(u)\partial_x \varphi) dt dx + \int_{\mathbb{R}} \eta(u_0(x))\varphi(0, x)dx \geq 0,$$

proving that u is the unique entropy solution of (7.2).

Finally, thanks to (7.6), (7.4) is proved.   □

7.2   Error Estimate

In this section, we complete the proof of Theorem 7.1 showing (7.5).

Let $t, \varepsilon > 0$. We "double the variables", using (τ, x) for (7.2) and (s, y) for (7.1). We have

$$
\begin{aligned}
&\partial_t |u(\tau, x) - u_\varepsilon(s, y)| \\
&\quad + \partial_x [\text{sign}\,(u(\tau, x) - u_\varepsilon(s, y))\,(f(u(\tau, x)) - f(u_\varepsilon(s, y)))] \leq 0,
\end{aligned}
\tag{7.10}
$$

and

$$
\begin{aligned}
&\partial_s |u(\tau, x) - u_\varepsilon(s, y)| \\
&\quad + \partial_y [\text{sign}\,(u(\tau, x) - u_\varepsilon(s, y))\,(f(u(\tau, x)) - f(u_\varepsilon(s, y)))] \\
&\leq \varepsilon \partial_{yy}^2 |u(\tau, x) - u_\varepsilon(s, y)|,
\end{aligned}
\tag{7.11}
$$

in the sense of distributions. Let $w \in C^\infty(\mathbb{R})$ be a nonnegative function with compact support such that

$$
\|w\|_{L^1(\mathbb{R})} = 1.
$$

We define

$$
w_\alpha(\xi) = \frac{1}{\alpha} w\left(\frac{\xi}{\alpha}\right), \qquad \xi \in \mathbb{R}, \ \alpha > 0.
$$

We test (7.10) with the function

$$
(\tau, x) \longmapsto w_\beta(\tau - s) w_\alpha(x - y), \qquad \alpha, \beta > 0,
$$

and we get

$$
\begin{aligned}
\int_0^t \int_\mathbb{R} \Big[&|u(\tau, x) - u_\varepsilon(s, y)| w_\beta'(\tau - s) w_\alpha(x - y) \\
&+ \text{sign}\,(u(\tau, x) - u_\varepsilon(s, y))\,(f(u(\tau, x)) - f(u_\varepsilon(s, y))) \cdot \\
&\qquad\qquad\qquad\qquad\qquad \cdot w_\beta(\tau - s) w_\alpha'(x - y) \Big] d\tau\, dx
\end{aligned}
$$

$$-\int_{\mathbb{R}}|u(t,x)-u_\varepsilon(s,y)|w_\beta(t-s)w_\alpha(x-y)dx$$

$$+\int_{\mathbb{R}}|u_0(x)-u_\varepsilon(s,y)|w_\beta(-s)w_\alpha(x-y)dx\geq 0,$$

that is

$$\int_0^t\int_{\mathbb{R}}\int_{\mathbb{R}}|u(t,x)-u_\varepsilon(s,y)|w_\beta(t-s)w_\alpha(x-y)dsdxdy$$

$$\leq\int_0^t\int_{\mathbb{R}}\int_{\mathbb{R}}|u_0(x)-u_\varepsilon(s,y)|w_\beta(-s)w_\alpha(x-y)dsdxdy$$

$$+\int_0^t\int_0^t\int_{\mathbb{R}}\int_{\mathbb{R}}\Big[|u(\tau,x)-u_\varepsilon(s,y)|w_\beta'(\tau-s)w_\alpha(x-y) \qquad (7.12)$$

$$+\operatorname{sign}(u(\tau,x)-u_\varepsilon(s,y))\,(f(u(\tau,x))-f(u_\varepsilon(s,y)))\cdot$$

$$\cdot w_\beta(\tau-s)w_\alpha'(x-y)\Big]dsd\tau dxdy.$$

We test (7.11) with the function

$$(s,y)\longmapsto w_\beta(\tau-s)w_\alpha(x-y), \qquad \alpha,\beta>0,$$

and we get

$$-\int_0^t\int_{\mathbb{R}}\Big[|u(\tau,x)-u_\varepsilon(s,y)|w_\beta'(\tau-s)w_\alpha(x-y)$$

$$+\operatorname{sign}(u(\tau,x)-u_\varepsilon(s,y))\,(f(u(\tau,x))-f(u_\varepsilon(s,y)))\cdot$$

$$\cdot w_\beta(\tau-s)w_\alpha'(x-y)\Big]dsdy$$

$$-\int_{\mathbb{R}}|u(\tau,x)-u_\varepsilon(t,y)|w_\beta(\tau-t)w_\alpha(x-y)dy$$

$$+\int_{\mathbb{R}}|u(\tau,x)-u_{0,\varepsilon}(y)|w_\beta(\tau)w_\alpha(x-y)dy$$

$$\geq-\varepsilon\int_0^t\int_{\mathbb{R}}|u(\tau,x)-u_\varepsilon(s,y)|w_\beta(\tau-s)w_\alpha''(x-y)dsdy,$$

that is

$$\int_0^t \int_{\mathbb{R}} \int_{\mathbb{R}} |u(\tau, x) - u_\varepsilon(t, y)| w_\beta(\tau - t) w_\alpha(x - y) d\tau dx dy$$

$$\leq \int_0^t \int_{\mathbb{R}} \int_{\mathbb{R}} |u(\tau, x) - u_{0,\varepsilon}(y)| w_\beta(\tau) w_\alpha(x - y) d\tau dx dy$$

$$- \int_0^t \int_0^t \int_{\mathbb{R}} \int_{\mathbb{R}} \Big[|u(\tau, x) - u_\varepsilon(s, y)| w_\beta'(\tau - s) w_\alpha(x - y)$$

$$+ \operatorname{sign}\left(u(\tau, x) - u_\varepsilon(s, y)\right) \left(f(u(\tau, x)) - f(u_\varepsilon(s, y))\right) \cdot$$

$$\cdot w_\beta(\tau - s) w_\alpha'(x - y) \Big] ds d\tau dx dy$$

$$+ \varepsilon \int_0^t \int_0^t \int_{\mathbb{R}} \int_{\mathbb{R}} |u(\tau, x) - u_\varepsilon(s, y)| w_\beta(\tau - s) w_\alpha''(x - y) ds d\tau dx dy.$$

$$(7.13)$$

We add (7.12) and (7.13),

$$\int_0^t \int_{\mathbb{R}} \int_{\mathbb{R}} |u(t, x) - u_\varepsilon(s, y)| w_\beta(t - s) w_\alpha(x - y) ds dx dy$$

$$+ \int_0^t \int_{\mathbb{R}} \int_{\mathbb{R}} |u(\tau, x) - u_\varepsilon(t, y)| w_\beta(\tau - t) w_\alpha(x - y) d\tau dx dy$$

$$\leq \int_0^t \int_{\mathbb{R}} \int_{\mathbb{R}} |u_0(x) - u_\varepsilon(s, y)| w_\beta(-s) w_\alpha(x - y) ds dx dy$$

$$+ \int_0^t \int_{\mathbb{R}} \int_{\mathbb{R}} |u(\tau, x) - u_{0,\varepsilon}(y)| w_\beta(\tau) w_\alpha(x - y) d\tau dx dy$$

$$+ \varepsilon \int_0^t \int_0^t \int_{\mathbb{R}} \int_{\mathbb{R}} |u(\tau, x) - u_\varepsilon(s, y)| w_\beta(\tau - s) w_\alpha''(x - y) ds d\tau dx dy,$$

and send $\beta \to 0$

$$\underbrace{\int_{\mathbb{R}} \int_{\mathbb{R}} |u(t, x) - u_\varepsilon(t, y)| w_\alpha(x - y) dx dy}_{I_1}$$

$$\leq \underbrace{\int_{\mathbb{R}} \int_{\mathbb{R}} |u_0(x) - u_{0,\varepsilon}(y)| w_\alpha(x - y) dx dy}_{I_2} \qquad (7.14)$$

$$+ \underbrace{\frac{\varepsilon}{2} \int_0^t \int_{\mathbb{R}} \int_{\mathbb{R}} |u(s, x) - u_\varepsilon(s, y)| w_\alpha''(x - y) ds dx dy}_{I_3}.$$

We estimate I_1 and I_2 in the following way (see (7.3) and Lemma 7.4)

$$I_1 \geq \int_{\mathbb{R}} \int_{\mathbb{R}} \Big(|u(t,x) - u_\varepsilon(t,x)| - |u_\varepsilon(t,x) - u_\varepsilon(t,y)| \Big) w_\alpha(x-y) dx dy$$

$$= \int_{\mathbb{R}} |u(t,x) - u_\varepsilon(t,x)| dx - \int_{\mathbb{R}} \int_{\mathbb{R}} |u_\varepsilon(t, y+\xi) - u_\varepsilon(t,y)| w_\alpha(\xi) d\xi dy$$

$$\geq \|u(t,\cdot) - u_\varepsilon(t,\cdot)\|_{L^1(\mathbb{R})} - \int_{\mathbb{R}} \left| \int_0^\xi \int_{\mathbb{R}} |\partial_x u_\varepsilon(t, y+\sigma)| dy d\sigma \right| w_\alpha(\xi) d\xi$$

$$= \|u(t,\cdot) - u_\varepsilon(t,\cdot)\|_{L^1(\mathbb{R})} - \|\partial_x u_\varepsilon(t,\cdot)\|_{L^1(\mathbb{R})} \int_{\mathbb{R}} |\xi| w_\alpha(\xi) d\xi$$

$$\geq \|u(t,\cdot) - u_\varepsilon(t,\cdot)\|_{L^1(\mathbb{R})} - \alpha TV(u_0) \int_{\mathbb{R}} |\xi| w(\xi) d\xi,$$

$$I_2 \leq \int_{\mathbb{R}} \int_{\mathbb{R}} \Big(|u_0(x) - u_{0,\varepsilon}(x)| + |u_{0,\varepsilon}(x) - u_{0,\varepsilon}(y)| \Big) w_\alpha(x-y) dx dy$$

$$= \int_{\mathbb{R}} |u_0(x) - u_{0,\varepsilon}(x)| dy + \int_{\mathbb{R}} \int_{\mathbb{R}} |u_{0,\varepsilon}(y+\xi) - u_{0,\varepsilon}(y)| w_\alpha(\xi) d\xi dy$$

$$\leq \|u_0 - u_{0,\varepsilon}\|_{L^1(\mathbb{R})} + \int_{\mathbb{R}} \left| \int_0^\xi \int_{\mathbb{R}} |\partial_x u_{0,\varepsilon}(y+\sigma)| dy d\sigma \right| w_\alpha(\xi) d\xi$$

$$= \|u_0 - u_{0,\varepsilon}\|_{L^1(\mathbb{R})} + \|\partial_x u_{0,\varepsilon}\|_{L^1(\mathbb{R})} \int_{\mathbb{R}} |\xi| w_\alpha(\xi) d\xi$$

$$\leq \|u_0 - u_{0,\varepsilon}\|_{L^1(\mathbb{R})} + \alpha TV(u_0) \int_{\mathbb{R}} |\xi| w(\xi) d\xi.$$

We have to estimate I_3. Thanks to (7.4), we know

$$I_3 = \lim_{\mu \to 0} I_{3,\mu},$$

where

$$I_{3,\mu} = \frac{\varepsilon}{2} \int_0^t \int_{\mathbb{R}} \int_{\mathbb{R}} |u_\mu(s,x) - u_\varepsilon(s,y)| w_\alpha''(x-y) ds dx dy, \qquad \mu > 0.$$

Since (see Lemma 7.4)

$$I_{3,\mu} \leq \frac{\varepsilon}{2} \int_0^t \int_{\mathbb{R}} \int_{\mathbb{R}} \Big(|\partial_x u_\mu(s,x)| + |\partial_y u_\varepsilon(s,y)| \Big) |w_\alpha'(x-y)| ds dx dy$$

$$= \frac{\varepsilon}{2} \int_0^t \int_{\mathbb{R}} \int_{\mathbb{R}} \Big(|\partial_x u_\mu(s, y+\xi)| + |\partial_y u_\varepsilon(s,y)| \Big) |w_\alpha'(\xi)| ds d\xi dy$$

$$= \frac{\varepsilon}{2} \int_0^t \int_{\mathbb{R}} (\|\partial_x u_\mu(s, \cdot)\|_{L^1(\mathbb{R})} + \|\partial_y u_\varepsilon(s, \cdot)\|_{L^1(\mathbb{R})}) |w'_\alpha(\xi)| ds d\xi$$

$$\leq \varepsilon t T V(u_0) \|w'_\alpha\|_{L^1(\mathbb{R})} = \frac{\varepsilon t}{\alpha} T V(u_0) \|w'\|_{L^1(\mathbb{R})},$$

we have

$$I_3 \leq \frac{\varepsilon t}{\alpha} T V(u_0) \|w'\|_{L^1(\mathbb{R})}.$$

Using the estimates on I_1, I_2, and I_3 in (7.14), we have

$$\|u(t, \cdot) - u_\varepsilon(t, \cdot)\|_{L^1(\mathbb{R})}$$

$$\leq \|u_0 - u_{0,\varepsilon}\|_{L^1(\mathbb{R})}$$

$$+ \left(\alpha + \frac{\varepsilon t}{\alpha}\right) T V(u_0) \left(2 \int_{\mathbb{R}} |\xi| w(\xi) d\xi + \|w'\|_{L^1(\mathbb{R})}\right).$$

Since the minimum of the map

$$\alpha \longmapsto \alpha + \frac{\varepsilon t}{\alpha}$$

is attained in $\sqrt{\varepsilon t}$, (7.5) is proved.

7.3   Exercises

Exercise 7.1 Let $n \in \mathbb{N} \setminus \{0\}$ and u be the solution of the Cauchy problem

$$\begin{cases} \partial_t u + \partial_x \log(u^2 + 1) = 0, & t > 0, x \in \mathbb{R}, \\ u(0, x) = -\frac{1}{x^2+1}, & x \in \mathbb{R}. \end{cases}$$

Prove that the function

$$t \longmapsto \int_{\mathbb{R}} u^n(t, x) dx$$

is nonincreasing if n is even and nondecreasing if n is odd.

References

1. Bardos, C., le Roux, A.Y., Nédélec, J.-C.: First order quasilinear equations with boundary conditions. Commun. Partial Differ. Equ. **4**(9), 1017–1034 (1979)
2. Bianchini, S., Bressan, A.: Vanishing viscosity solutions of nonlinear hyperbolic systems. Ann. Math. (2) **161**(1), 223–342 (2005)
3. Holden, H., Risebro, N.H.: Front Tracking for Hyperbolic Conservation Laws. Applied Mathematical Sciences, vol. 152, 2nd edn. Springer, Heidelberg (2015)
4. Kružkov, S.N.: First order quasilinear equations with several independent variables. Mat. Sb. (N.S.) **81**(123), 228–255 (1970)
5. Kuznecov, N.N.: The accuracy of certain approximate methods for the computation of weak solutions of a first order quasilinear equation. Ž. Vyčisl. Mat. i Mat. Fiz. **16**(6), 1489–1502 (1627/1976)
6. Serre, D.: Systems of Conservation Laws, vol. 1. Cambridge University Press, Cambridge (1999). Hyperbolicity, entropies, shock waves, Translated from the 1996 French original by I. N. Sneddon

Chapter 8
Compensated Compactness

Abstract In this chapter we present the compensated compactness approach to the existence issue. It is a useful measure theory tool based on the compression effects of nonlinear equations. It works only in the nonlinear case and applies in several convergence problems.

Keywords Compensated compactness · Existence · Genuine nonlinearity · Radon measures · Young measures · Murat lemma · Div-Curl lemma · Bounded solutions

In this chapter, using a compensated compactness argument, we prove the existence of entropy weak solutions for the Cauchy problem

$$\begin{cases} \partial_t u + \partial_x f(u) = 0, & t > 0, \ x \in \mathbb{R}, \\ u(0, x) = u_0(x), & x \in \mathbb{R}, \end{cases} \tag{8.1}$$

when

$$u_0 \in L^2(\mathbb{R}) \cap L^\infty(\mathbb{R}), \qquad |\{f'' = 0\}| = 0,$$

in this case we say that f is *genuinely nonlinear*. Using a vanishing viscosity approximation of (8.1), we can find a family of approximate solutions $\{u_\varepsilon\}_{\varepsilon>0}$. In general, $\{u_\varepsilon\}_{\varepsilon>0}$ is bounded in $L^\infty((0, \infty) \times \mathbb{R})$ but not in $BV_{loc}((0, \infty) \times \mathbb{R})$. Since also $\{f(u_\varepsilon)\}_{\varepsilon>0}$ is bounded in $L^\infty((0, \infty) \times \mathbb{R})$, there exist a sequence $\{u_{\varepsilon_k}\}_{k\in\mathbb{N}}$, $\varepsilon_k \to 0$, and two functions $u, \overline{f} \in L^\infty((0, \infty) \times \mathbb{R})$ such that

$$u_{\varepsilon_k} \overset{\star}{\rightharpoonup} u, \quad f(u_{\varepsilon_k}) \overset{\star}{\rightharpoonup} \overline{f}, \qquad \text{weakly-} * \text{ in } L^\infty((0, \infty) \times \mathbb{R}). \tag{8.2}$$

Since $(L^1)' = L^\infty$, (8.2), means

$$\forall \varphi \in L^1((0,\infty) \times \mathbb{R}) : \quad \begin{aligned} \int_0^\infty \int_{\mathbb{R}} u_{\varepsilon_k} \varphi dx &\longrightarrow \int_0^\infty \int_{\mathbb{R}} u\varphi dx, \\ \int_0^\infty \int_{\mathbb{R}} f(u_{\varepsilon_k})\varphi dx &\longrightarrow \int_0^\infty \int_{\mathbb{R}} \overline{f}\varphi dx. \end{aligned}$$

Unfortunately, in general

$$f(u) \neq \overline{f},$$

in particular, if f is convex, we can only say

$$f(u) \leq \overline{f}, \qquad \text{a.e. in } [0,\infty) \times \mathbb{R}.$$

Clearly, we cannot deduce that u is an entropy weak solution of (8.1).

The compensated compactness [4, Chapter XVI] is a tool in measure theory that, using the structure of the equation in (8.1), gives us the strong convergence of $\{u_{\varepsilon_k}\}_{k \in \mathbb{N}}$ and then of $\{f(u_{\varepsilon_k})\}_{k \in \mathbb{N}}$.

8.1　Young Measures

Let $\Omega \subset \mathbb{R}^N$, $N \geq 2$, be an open measurable set and μ be a measure on Ω.

Definition 8.1 We say that μ is a Radon measure on Ω and we write

$$\mu \in \mathcal{M}(\Omega),$$

if

(i) μ is a Borel measure;
(ii) μ is locally finite, namely every point of Ω has a neighborhood with finite measure;
(iii) μ is inner regular, namely for every measurable $E \subset \Omega$

$$\mu(E) = \sup\{\mu(K); K \subset E, K \text{ compact}\};$$

(iv) μ is outer regular, namely for every measurable $E \subset \Omega$

$$\mu(E) = \inf\{\mu(U); E \subset U, U \text{ open}\}.$$

We remind that the Riesz' representation theorem says [7, Theorem 2.14]

$$M(\Omega) = \left(C(\Omega) \cup L^{\infty}(\Omega)\right)'. \tag{8.3}$$

Moreover, we have the duality formula [2]

$$L_w^{\infty}(\Omega; M(\mathbb{R})) = \left(L^1(\Omega; C(\mathbb{R}) \cap L^{\infty}(\mathbb{R}))\right)', \tag{8.4}$$

where $L_w^{\infty}(\Omega; M(\mathbb{R}))$ is the space of all the essentially bounded weak-∗ measurable maps from Ω in $M(\mathbb{R})$, namely

$$(\mu_x)_{x \in \Omega} \in L_w^{\infty}(\Omega; M(\mathbb{R})), \tag{8.5}$$

if and only if

(i) for every $x \in \Omega$, $\mu_x \in M(\mathbb{R})$;
(ii) ess $\sup_{x \in \Omega} |\mu_x|(\mathbb{R}) < \infty$;
(iii) for every $g \in C(\mathbb{R}) \cap L^{\infty}(\mathbb{R})$ the map $x \mapsto \int_{\mathbb{R}} g(y) d\mu_x(y)$ is measurable.[1]

The main result of this section is the following.

Theorem 8.1 *Let $\Omega \subset \mathbb{R}^N$, $N \geq 2$, be open and $\{u_n\}_{n \in \mathbb{N}}$ a sequence of function defined on Ω with values in $\mathbb{R}$. If*

$$\{u_n\}_{n \in \mathbb{N}} \text{ is uniformly bounded in } L^{\infty}(\Omega),$$

then there exist

(1) a subsequence $\{u_{n_k}\}_{k \in \mathbb{N}}$ of $\{u_n\}_{n \in \mathbb{N}}$;
(2) a measurable family of probability measures on $\mathbb{R}$ $\{v_x\}_{x \in \Omega}$ with compact support;[2]

such that, for every $g \in C(\mathbb{R}) \cap L^{\infty}(\mathbb{R})$, we have that

$$g(u_{n_k}) \overset{\star}{\rightharpoonup} \overline{g}, \qquad \text{weakly-} \ast \text{ in } L^{\infty}(\Omega), \tag{8.6}$$

where

$$\overline{g}(x) = \int_{\mathbb{R}} g(y) dv_x(y), \qquad x \in \Omega, \tag{8.7}$$

---

[1] This is the definition of *weak-∗ measurability*.
[2] A family of measures on $\mathbb{R}$ $\{v_x\}_{x \in \Omega}$ is termed *Young measure* on $\mathbb{R}$.

namely

$$\forall \varphi \in L^1(\Omega) : \int_\Omega g(u_{n_k})\varphi dx \longrightarrow \int_\Omega \int_\mathbb{R} g(y)\varphi(x)dv_x(y)dx.$$

Proof Let $n \in \mathbb{N}$. Consider the functions

$$v_n : x \in \Omega \longmapsto \delta_{u_n(x)} \in \mathcal{M}(\mathbb{R}),$$

where $\delta_{u_n(x)}$ is the Dirac's delta concentrated on the singleton $\{u_n(x)\}$. We have that

$$v_n \in L^\infty(\Omega; \mathcal{M}(\mathbb{R})),$$

and, in addition,

$$\{v_n\}_{n\in\mathbb{N}} \text{ is bounded in } L^\infty(\Omega; \mathcal{M}(\mathbb{R})).$$

Therefore, the Banach-Alaoglou theorem [3, Theorem 3.16] says that there exist a subsequence $\{v_{n_k}\}_{k\in\mathbb{N}}$ of $\{v_n\}_{n\in\mathbb{N}}$ and a function $v \in L^\infty(\Omega; \mathcal{M}(\mathbb{R}))$ such that

$$v_{n_k} \overset{*}{\rightharpoonup} v, \qquad \text{weakly-} * \text{ in } L^\infty(\Omega; \mathcal{M}(\mathbb{R})). \tag{8.8}$$

By identifying

$$v = \{v_x\}_{x\in\Omega},$$

(8.8) tells that for every $\Psi \in L^1(\Omega; C(\mathbb{R}) \cap L^\infty(\mathbb{R}))$

$$\int_\Omega \Psi(x, u_{n_k}(x))dx = \int_\Omega \int_\mathbb{R} \Psi(x, y)d\delta_{u_{n_k}(x)}(y)dx$$

$$= \int_\Omega \int_\mathbb{R} \Psi(x, y)dv_{n_k}(y)dx \longrightarrow \int_\Omega \int_\mathbb{R} \Psi(x, y)dv_x(y)dx.$$

We claim that v_x is a probability measure for almost every $x \in \Omega$. Let $\varphi \in L^1(\Omega)$. Consider a a function $\Psi \in L^1(\Omega; C(\mathbb{R}) \cap L^\infty(\mathbb{R}))$ such that

$$\Psi(x, \cdot) = \varphi(x), \qquad \forall x \in \Omega, \tag{8.9}$$

we have

$$\int_\Omega \varphi(x)dx = \int_\Omega \Psi(x, u_{n_k}(x))dx$$

$$= \int_\Omega \int_\mathbb{R} \Psi(x, y)d\delta_{u_{n_k}(x)}(y)dx, \qquad k \in \mathbb{N}.$$

As $k \to \infty$

$$\int_{\Omega} \varphi(x)dx = \int_{\Omega} \int_{\mathbb{R}} \Psi(x, y)d\nu_x(y)dx,$$

and, thanks to (8.9),

$$\int_{\Omega} \varphi(x)dx = \int_{\Omega} \int_{\mathbb{R}} \varphi(x)d\nu_x(y)dx = \int_{\Omega} \varphi(x) \left(\int_{\mathbb{R}} d\nu_x(y) \right) dx.$$

Therefore

$$\int_{\mathbb{R}} d\nu_x(y) = 1, \qquad \text{a.e. } x \in \Omega.$$

Since

$$\text{supp}(\delta_{u_{n_k}(x)}) \subset \left[-\sup_n \|u_n\|_{L^{\infty}(\Omega)}, \sup_n \|u_n\|_{L^{\infty}(\Omega)} \right], \qquad k \in \mathbb{N}, \ x \in \Omega,$$

the measures ν_x, $x \in \Omega$, have compact support.

We conclude the proof, showing that (8.6) holds. Let $\varphi \in L^1(\Omega)$ and $g \in C(\mathbb{R}) \cap L^{\infty}(\mathbb{R})$. Since the function

$$(x, y) \in \Omega \times \mathbb{R} \longmapsto \varphi(x)g(y)$$

belongs to $L^1(\Omega; C(\mathbb{R}) \cap L^{\infty}(\mathbb{R}))$, from (8.8), we have that

$$\int_{\Omega} \varphi(x)g(u_{n_k}(x))dx \longrightarrow \int_{\Omega} \int_{\mathbb{R}} \varphi(x)g(y)d\nu_x(y)dx = \int_{\Omega} \varphi\overline{g}dx,$$

where $\overline{g}$ is defined in (8.7). Therefore, (8.6) holds and the proof is concluded.   □

8.2   Murat Compact Embedding and Div-Curl Lemma

We remind that for every open bounded set $\Omega \subset \mathbb{R}^N$, $N \geq 2$, we have

$$H^{-1}(\Omega) = \left(H_0^1(\Omega) \right)',$$

namely

$$f \in H^{-1}(\Omega)$$

$$\Updownarrow$$

there exists $C > 0$ s.t. for all $\varphi \in H_0^1(\Omega)$: $\left| \int_\Omega f \varphi dx \right| \leq C \|\varphi\|_{H_0^1(\Omega)}$.

On the compact subsets of $H^{-1}(\Omega)$ we have the following fundamental Murat result [6].

Theorem 8.2 (Murat Lemma) *Let $\Omega \subset \mathbb{R}^N$, $N \geq 2$, be open and bounded, $\{\varphi_n\}_{n \in \mathbb{N}}$, $\{\psi_n\}_{n \in \mathbb{N}}$, and $\{\chi_n\}_{n \in \mathbb{N}}$ be three sequences of distributions defined on Ω, and $2 < p \leq \infty$. If*

(i) $\varphi_n = \psi_n + \chi_n$, for all $n \in \mathbb{N}$;
(ii) $\{\varphi_n\}_{n \in \mathbb{N}}$ is bounded in $W^{-1,p}(\Omega)$;
(iii) $\{\psi_n\}_{n \in \mathbb{N}}$ is bounded in $\mathcal{M}(\Omega)$;
(iv) $\{\chi_n\}_{n \in \mathbb{N}}$ is compact in $H^{-1}(\Omega)$;

then $\{\varphi_n\}_{n \in \mathbb{N}}$ is compact in $H^{-1}(\Omega)$.

The proof of this theorem is based on the following fundamental results on elliptic equations and Sobolev spaces.

Theorem 8.3 (Agmon [1]) *Let $\Omega \subset \mathbb{R}^N$, $N \geq 2$, be open and bounded, $1 < p \leq \infty$, and $\varphi \in L^p(\Omega)$. The elliptic homogeneous Dirichlet problem*

$$\begin{cases} -\Delta u = \varphi, & in \ \Omega, \\ u = 0, & on \ \partial\Omega, \end{cases} \tag{8.10}$$

admits a unique distributional solution u. Moreover,

$$u \in W_0^{1,p}(\Omega) \cap W^{2,p}(\Omega).$$

Theorem 8.4 (Stampacchia [8]) *Let $\Omega \subset \mathbb{R}^N$, $N \geq 2$, be open and bounded, and $\varphi \in L^1(\Omega)$. The elliptic homogeneous Dirichlet problem*

$$\begin{cases} -\Delta u = \varphi, & in \ \Omega, \\ u = 0, & on \ \partial\Omega, \end{cases} \tag{8.11}$$

admits a unique distributional solution u. Moreover,

$$u \in W_0^{1,q}(\Omega), \qquad 1 \leq q < \frac{N}{N-1}.$$

Theorem 8.5 (Rellich-Kondrakov [5, §5.7]) *Let $\Omega \subset \mathbb{R}^N$, $N \geq 2$, be open and bounded, and $1 < p \leq \infty$. The following compact embeddings hold*

$$W^{1,p}(\Omega) \hookrightarrow\hookrightarrow \begin{cases} L^q(\Omega), \ 1 \leq q < Np/(N-p), & \text{if } 1 \leq p < N, \\ L^q(\Omega), \ 1 \leq q < \infty, & \text{if } p = N, \\ C(\overline{\Omega}), & \text{if } p > N. \end{cases}$$

Proof of Theorem 8.2 Let $n \in \mathbb{N}$ and $u_n \in W_0^{1,p}(\Omega)$, $v_n \in W_0^{1,q}(\Omega)$, $1 \leq q < \frac{N}{N-1}$, $w_n \in H_0^1(\Omega)$ be the solutions of the elliptic homogeneous Dirichlet problems (see Theorems 8.3 and 8.4)

$$\begin{cases} -\Delta u_n = \varphi_n, & \text{in } \Omega, \\ u_n = 0, & \text{on } \partial\Omega, \end{cases}$$

$$\begin{cases} -\Delta v_n = \psi_n, & \text{in } \Omega, \\ v_n = 0, & \text{on } \partial\Omega, \end{cases}$$

$$\begin{cases} -\Delta w_n = \chi_n, & \text{in } \Omega, \\ w_n = 0, & \text{on } \partial\Omega, \end{cases}$$

respectively. Due to our assumptions we have that

(1) $u_n = v_n + w_n$, for all $n \in \mathbb{N}$;
(2) $\{u_n\}_{n\in\mathbb{N}}$ is bounded in $W_0^{1,p}(\Omega)$;
(3) since $\mathcal{M}(\Omega)$ is locally compact, $\{\psi_n\}_{n\in\mathbb{N}}$ is compact in $\mathcal{M}(\Omega)$, then $\{v_n\}_{n\in\mathbb{N}}$ is compact in $W_0^{1,q}(\Omega)$, $1 \leq q < \frac{N}{N-1}$;
(4) $\{w_n\}_{n\in\mathbb{N}}$ is compact in $H_0^1(\Omega)$.

As a consequence $\{u_n\}_{n\in\mathbb{N}}$ is bounded in $W_0^{1,p}(\Omega)$ and compact in $W_0^{1,q}(\Omega)$, $1 \leq q < \frac{N}{N-1}$. Since

$$1 \leq q < \frac{N}{N-1} < 2 < p < \infty,$$

the interpolation theorems guarantee the compactness of $\{u_n\}_{n\in\mathbb{N}}$ in $W_0^{1,q}(\Omega)$, $1 \leq q < p$. In particular $\{u_n\}_{n\in\mathbb{N}}$ is compact in $H_0^1(\Omega)$ and then $\{\psi_n\}_{n\in\mathbb{N}}$ is compact in $H^{-1}(\Omega)$. □

We are now ready for the Div-Curl Lemma.

Theorem 8.6 (Div-Curl Lemma) *Let $\Omega \subset \mathbb{R}^N$, $N \geq 2$, be open, $\{G_n\}_{n\in\mathbb{N}}$, $\{H_n\}_{n\in\mathbb{N}} \subset L^2(\Omega; \mathbb{R}^N)$ and $G, H \in L^2(\Omega; \mathbb{R}^N)$. If*

(i) $G_n \rightharpoonup G$, $H_n \rightharpoonup H$ weakly in $L^2(\Omega; \mathbb{R}^N)$,

(ii) $\{\operatorname{div}(G_n)\}_{n \in \mathbb{N}}$, $\{\operatorname{curl}(H_n)\}_{n \in \mathbb{N}}$ are compact in $H^{-1}(\Omega)$,

then

$$G_n \cdot H_n \longrightarrow G \cdot H \qquad \text{in the sense of distributions in } \Omega,$$

namely

$$\forall \varphi \in C_c^\infty(\Omega) : \int_\Omega G_n \cdot H_n \varphi \, dx \longrightarrow \int_\Omega G \cdot H \varphi \, dx.$$

Proof It is not restrictive to assume that

$$H \equiv 0.$$

Let $n \in \mathbb{N}$ and $\Phi_n \in H^2(\Omega; \mathbb{R}^N) \cap H_0^1(\Omega; \mathbb{R}^N)$ be the solution of the elliptic homogeneous Dirichlet problem (see Theorem 8.3)

$$\begin{cases} -\Delta \Phi_n = H_n, & \text{in } \Omega, \\ \Phi_n = 0, & \text{on } \partial\Omega. \end{cases}$$

Since

$$-\Delta \operatorname{div}(\Phi_n) = \operatorname{div}(H_n), \qquad -\Delta \operatorname{curl}(\Phi_n) = \operatorname{curl}(H_n),$$

we have that

(1) $\Phi_n \rightharpoonup 0$ weakly in $H_{loc}^2(\Omega; \mathbb{R}^N)$;
(2) $\operatorname{div}(\Phi_n) \rightharpoonup 0$ weakly in $H_{loc}^1(\Omega; \mathbb{R}^N)$;
(3) $\{\operatorname{curl}(\Phi_n)\}_{n \in \mathbb{N}}$ is compact in $H_{loc}^1(\Omega; \mathbb{R}^N)$, therefore, passing to a subsequence,

$$\operatorname{curl}(\Phi_n) \longrightarrow 0 \text{ in } H_{loc}^1(\Omega; \mathbb{R}^N).$$

Writing

$$V_n := H_n + \nabla \operatorname{div}(\Phi_n),$$

we have that

$$\begin{aligned} V_n &= H_n + \nabla \operatorname{div}(\Phi_n) \\ &= -\Delta \Phi_n + \nabla \operatorname{div}(\Phi_n) \\ &= -\operatorname{div}(\nabla \Phi_n) + \nabla \operatorname{div}(\Phi_n) = \operatorname{curl}(\operatorname{curl}(\Phi_n)), \end{aligned}$$

therefore, passing to a subsequence, (see (3))

$$V_n \longrightarrow 0 \text{ in } L^2_{loc}(\Omega; \mathbb{R}^N). \tag{8.12}$$

Moreover, we observe that

$$G_n \cdot H_n = G_n \cdot \left(V_n - \nabla \text{div}(\Phi_n) \right)$$

$$= G_n \cdot V_n - G_n \cdot \nabla \text{div}(\Phi_n)$$

$$= G_n \cdot V_n - \text{div}(G_n \text{div}(\Phi_n)) + \text{div}(G_n) \text{div}(\Phi_n).$$

Since, in light of (i), (ii), (2), and (8.12)

$$G_n \cdot V_n \longrightarrow 0, \quad G_n \text{div}(\Phi_n) \longrightarrow 0, \quad \text{div}(G_n) \text{div}(\Phi_n) \longrightarrow 0,$$

in the sense of distributions in Ω, the claim is proved.   □

8.3 Scalar Conservation Laws

The main result of this section is the following theorem [9, 10].

Theorem 8.7 *Let $\Omega \subset [0, \infty) \times \mathbb{R}$ be open, $f \in C^2(\mathbb{R})$, and $\{u_n\}_{n \in \mathbb{N}}$ be a sequence of functions defined in Ω with values in $\mathbb{R}$. If*

(i) $\{u_n\}_{n \in \mathbb{N}}$ is uniformly bounded in $L^\infty(\Omega)$;
(ii) for every entropy $\eta \in C^2(\mathbb{R})$ with entropy flux $q \in C^2(\mathbb{R})$ defined by $q' = f'\eta'$, the sequence $\{\partial_t \eta(u_n) + \partial_x q(u_n)\}_{n \in \mathbb{N}}$ is compact in $H^{-1}(\Omega)$;
(iii) f is genuinely nonlinear, i.e., $|\{f'' = 0\}| = 0$;

then, there exist a subsequence $\{u_{n_k}\}_{k \in \mathbb{N}}$ and a function $u \in L^\infty(\Omega)$ such that

$$u_{n_k} \longrightarrow u \quad \text{in } L^p_{loc}(\Omega), \ 1 \leq p < \infty, \ \text{and a.e. in } \Omega.$$

Proof Thanks to Theorem 8.1 there exist a subsequence $\{u_{n_k}\}_{k \in \mathbb{N}}$ of $\{u_n\}_{n \in \mathbb{N}}$ and a measurable family of probability measures on $\mathbb{R}$ $\{v_{(t,x)}\}_{(t,x) \in \Omega}$ with compact support such that, for every $g \in C(\mathbb{R}) \cap L^\infty(\mathbb{R})$, we have that

$$g(u_{n_k}) \overset{\star}{\rightharpoonup} \overline{g}, \quad \text{weakly-}* \text{ in } L^\infty(\Omega), \tag{8.13}$$

where

$$\overline{g}(t, x) = \int_{\mathbb{R}} g(y) dv_{(t,x)}(y), \quad (t, x) \in \Omega.$$

Due to the uniform boundedness of $\{u_n\}_n$ in $L^\infty(\Omega)$ and the compact support of $v_{(t,x)}$, (8.13) holds for every continuous function.

We consider the function

$$g(\xi) = \xi, \qquad \xi \in \mathbb{R},$$

and introduce the notation

$$u = \overline{g}.$$

Therefore, (8.13) says

$$u_{n_k} \overset{\star}{\rightharpoonup} u, \qquad \text{weakly-} * \text{ in } L^\infty(\Omega). \tag{8.14}$$

Step I.   We claim that

$$f(u) = \overline{f}. \tag{8.15}$$

Consider the entropies

$$\eta_1(\xi) = g(\xi) = \xi, \qquad \eta_2(\xi) = f(\xi), \qquad \xi \in \mathbb{R},$$

with fluxes

$$q_1(\xi) = f(\xi), \qquad q_2(\xi) = \int_0^\xi (f'(s))^2 ds, \qquad \xi \in \mathbb{R},$$

respectively. We prove that

$$u\overline{q_2} - \overline{f}^2 = \overline{gq_2 - f^2}. \tag{8.16}$$

Introducing the vector fields

$$G_{n_k} = \left(\eta_1(u_{n_k}), q_1(u_{n_k})\right), \qquad H_{n_k} = \left(q_2(u_{n_k}), -\eta_2(u_{n_k})\right),$$

we have that

$$G_{n_k}, \; H_{n_k} \in L^2(\Omega; \mathbb{R}^2),$$

$$G_{n_k} \rightharpoonup \overline{G} = (\overline{\eta_1}, \overline{q_1}), \; H_{n_k} \rightharpoonup \overline{H} = (\overline{q_2}, -\overline{\eta_2}), \quad \text{weakly in } L^2(\Omega; \mathbb{R}^2),$$

$$\operatorname{div}\left(G_{n_k}\right) = \partial_t \eta_1(u_{n_k}) + \partial_x q_1(u_{n_k}),$$

$$\operatorname{curl}\left(H_{n_k}\right) = \partial_t \eta_2(u_{n_k}) + \partial_x q_2(u_{n_k}),$$

therefore $\{\text{div}\left(G_{n_k}\right)\}_{n_k}$ and $\{\text{curl}\left(H_{n_k}\right)\}_{n_k}$ are compact in $H^{-1}(\Omega)$. Theorem 8.6 says

$$\eta_1(u_{n_k})q_2(u_{n_k}) - \eta_2(u_{n_k})q_1(u_{n_k})$$
$$\longrightarrow \overline{\eta_1}\,\overline{q_2} - \overline{\eta_2}\,\overline{q_1}, \quad \text{in the sense of distributions in } \Omega,$$

on the other hand

$$\eta_1(u_{n_k})q_2(u_{n_k}) - \eta_2(u_{n_k})q_1(u_{n_k}) \overset{\star}{\rightharpoonup} \overline{\eta_1 q_2 - \eta_2 q_1}, \quad \text{weakly-} * \text{ in } L^\infty(\Omega),$$

therefore

$$\overline{\eta_1}\,\overline{q_2} - \overline{\eta_2}\,\overline{q_1} = \overline{\eta_1 q_2 - \eta_2 q_1},$$

that is (8.16).
The following inequality holds

$$(f(\xi) - f(\sigma))^2 \le (\xi - \sigma)\,(q_2(\xi) - q_2(\sigma)), \qquad \xi, \sigma \in \mathbb{R}. \qquad (8.17)$$

Indeed, if $\xi \le \sigma$, using the Hölder inequality,

$$(f(\xi) - f(\sigma))^2 = \left(\int_\xi^\sigma f'(\tau)d\tau\right)^2$$

$$\le (\sigma - \xi)\int_\xi^\sigma \left(f'(\tau)\right)^2 d\tau = (\sigma - \xi)\,(q_2(\sigma) - q_2(\xi)),$$

and when $\xi \ge \sigma$ one can argue in the same way.
We have that

$$\left(f(u_{n_k}) - f(u)\right)^2 - (u_{n_k} - u)\left(q_2(u_{n_k}) - q_2(u)\right) \overset{\star}{\rightharpoonup}$$
$$\overset{\star}{\rightharpoonup} \left(\overline{f} - f(u)\right)^2, \qquad \text{weakly-} * \text{ in } L^\infty(\Omega). \qquad (8.18)$$

Indeed

$$\left(f(u_{n_k}) - f(u)\right)^2 - (u_{n_k} - u)\left(q_2(u_{n_k}) - q_2(u)\right)$$
$$= \left(f^2(u_{n_k}) - u_{n_k}q_2(u_{n_k})\right) - 2f(u_{n_k})f(u)$$
$$+ f^2(u) + u_{n_k}q_2(u) + uq_2(u_{n_k}) - uq_2(u),$$

as $k \to \infty$, due to (8.16),

$$\overline{f^2 - g q_2} - 2\overline{f} f(u) + f^2(u) + \overline{u q_2}(u) + u\overline{q_2} - u q_2(u)$$
$$= \overline{f}^2 - u\overline{q_2} - 2\overline{f} f(u) + f^2(u) + u\overline{q_2} = \left(\overline{f} - f(u)\right)^2 .$$

Thanks to (8.17)

$$\left(f(u_{n_k}) - f(u)\right)^2 - (u_{n_k} - u)\left(q_2(u_{n_k}) - q_2(u)\right) \leq 0,$$

therefore, thanks to (8.18),

$$\left(\overline{f} - f(u)\right)^2 \leq 0,$$

that gives (8.15).
Using (8.15) in (8.18), we have also

$$\left(f(u_{n_k}) - f(u)\right)^2 - (u_{n_k} - u)\left(q_2(u_{n_k}) - q_2(u)\right) \overset{\star}{\rightharpoonup} 0,$$

$$\text{weakly-} * \text{ in } L^\infty(\Omega).$$
$$(8.19)$$

Step II.   Let us prove that

$$\nu_{(t,x)} = \delta_{u(t,x)}, \qquad (t, x) \in \Omega. \tag{8.20}$$

(8.19) says

$$\int_{\mathbb{R}} \left[(f(y) - f(u(t, x)))^2 - (y - u(t, x))\,(q_2(y) - q_2(u(t, x))) \right] d\nu_{(t,x)}(y) = 0,$$

therefore, thanks to (8.17),

$$(f(y) - f(u(t, x)))^2$$
$$= (y - u(t, x))\,(q_2(y) - q_2(u(t, x))), \tag{8.21}$$

for every $(t, x) \in \Omega$, $y \in \text{supp}(\nu_{(t,x)})$. Let $(t, x) \in \Omega$, $y \in \text{supp}(\nu_{(t,x)})$. It is not restrictive to consider only the case

$$u(t, x) \leq y.$$

Assume by contradiction that

$$u(t, x) < y.$$

The definition of q_2 and (8.21) give

$$0 = (f(y) - f(u(t,x)))^2 - (y - u(t,x)) (q_2(y) - q_2(u(t,x)))$$

$$= \left(\int_{u(t,x)}^y f'(s)ds \right)^2 - (y - u(t,x)) \int_{u(t,x)}^y (f'(s))^2 ds$$

$$= (y - u(t,x))^2 \left(\left(\frac{1}{y - u(t,x)} \int_{u(t,x)}^y f'(s)ds \right)^2 \right.$$

$$\left. - \frac{1}{y - u(t,x)} \int_{u(t,x)}^y (f'(s))^2 ds \right),$$

that is

$$\left(\frac{1}{y - u(t,x)} \int_{u(t,x)}^y f'(s)ds \right)^2 = \frac{1}{y - u(t,x)} \int_{u(t,x)}^y (f'(s))^2 ds. \qquad (8.22)$$

Due to the strict convexity of the map $\xi \mapsto \xi^2$, the Jensen inequality gives

$$\left(\frac{1}{y - u(t,x)} \int_{u(t,x)}^y f'(s)ds \right)^2 \leq \frac{1}{y - u(t,x)} \int_{u(t,x)}^y (f'(s))^2 ds.$$

Since in (8.22) we have the equality, f' must be constant in $[u(t,x), y]$, and then

$$f''(\xi) = 0, \qquad \forall \xi \in [u(t,x), y],$$

that contradicts (iii).
Therefore we must have

$$u(t,x) = y.$$

As a consequence (8.20) holds.
Step III. We conclude by proving

$$u_{n_k} \longrightarrow u, \qquad \text{a.e. and in } L_{loc}^p(\Omega), \ 1 \leq p < \infty. \qquad (8.23)$$

Let $g \in C(\mathbb{R})$, due to (8.21),

$$\overline{g}(t,x) = \int_{\mathbb{R}} g(y)d\nu_{(t,x)}(y) = \int_{\mathbb{R}} g(y)d\delta_{u(t,x)}(y) = g(u(t,x)),$$

therefore

$$g(u_{n_k}) \overset{\star}{\rightharpoonup} g(u), \qquad \text{weakly-}* \text{ in } L^\infty(\Omega),$$

and in particular

$$u_{n_k}^2 \overset{\star}{\rightharpoonup} u^2, \qquad \text{weakly-} * \text{ in } L^\infty(\Omega).$$

For every $K \subset \Omega$ compact we have

$$\int_K (u_{n_k} - u)^2 dx = \int_K u_{n_k}^2 dx + \int_K u^2 dx - 2\int_K u_{n_k} u \, dx \longrightarrow 0,$$

that is

$$u_{n_k} \longrightarrow u, \qquad \text{in } L^2_{loc}(\Omega).$$

Therefore the uniform boundedness in L^∞ gives (8.23).

□

8.4   Global Existence of Bounded Solutions

We finally prove the following existence result.

Theorem 8.8 *Let $f \in C^2(\mathbb{R})$ and $u_0 : \mathbb{R} \to \mathbb{R}$. If f is genuinely nonlinear and $u_0 \in L^2(\mathbb{R}) \cap L^\infty(\mathbb{R})$, then (8.1) has an entropy weak solution $u \in L^\infty((0, \infty) \times \mathbb{R})$.*

Proof Let $\varepsilon > 0$ and

$$u_\varepsilon \in C^\infty([0, \infty) \times \mathbb{R}) \cap W^{2,p}((0, \infty); W^{1,p}(\mathbb{R})), \qquad 1 \le p < \infty,$$

be the solution of the parabolic problem

$$\begin{cases} \partial_t u_\varepsilon + \partial_x f(u_\varepsilon) = \varepsilon \partial_{xx}^2 u_\varepsilon, & t > 0, \ x \in \mathbb{R}, \\ u_\varepsilon(0, x) = u_{0,\varepsilon}(x), & x \in \mathbb{R}, \end{cases} \qquad (8.24)$$

where $u_{0,\varepsilon}$ is a smooth approximation of u_0 such that

$$u_{0,\varepsilon} \in C^\infty(\mathbb{R}) \cap W^{2,1}(\mathbb{R}), \qquad \varepsilon > 0,$$

$$u_{0,\varepsilon} \longrightarrow u_0, \qquad \text{in } L^p_{loc}(\mathbb{R}), \ 1 \le p < \infty, \text{ as } \varepsilon \to 0, \qquad (8.25)$$

$$\left\| u_{0,\varepsilon} \right\|_{L^\infty(\mathbb{R})} \le \left\| u_0 \right\|_{L^\infty(\mathbb{R})}, \quad \left\| u_{0,\varepsilon} \right\|_{L^2(\mathbb{R})} \le \left\| u_0 \right\|_{L^2(\mathbb{R})}, \qquad \varepsilon > 0.$$

The maximum principle for parabolic equations says

$$\left\| u_\varepsilon \right\|_{L^\infty((0,\infty) \times \mathbb{R})} \le \left\| u_0 \right\|_{L^\infty(\mathbb{R})}, \qquad \varepsilon > 0. \qquad (8.26)$$

Let us prove the energy estimate

$$\|u_\varepsilon(t,\cdot)\|^2_{L^2(\mathbb{R})} + 2\varepsilon \int_0^t \|\partial_x u_\varepsilon(s,\cdot)\|^2_{L^2(\mathbb{R})}\,ds \le \|u_0\|^2_{L^2(\mathbb{R})}, \qquad t \ge 0, \ \varepsilon > 0.$$

(8.27)

We have that

$$\frac{d}{dt}\int_{\mathbb{R}} \frac{u_\varepsilon^2}{2}dx = \int_{\mathbb{R}} u_\varepsilon \partial_t u_\varepsilon dx$$

$$= \varepsilon \int_{\mathbb{R}} u_\varepsilon \partial_{xx}^2 u_\varepsilon dx - \int_{\mathbb{R}} u_\varepsilon f'(u_\varepsilon)\partial_x u_\varepsilon dx$$

$$= -\varepsilon \int_{\mathbb{R}} (\partial_x u_\varepsilon)^2\,dx - \underbrace{\int_{\mathbb{R}} \partial_x \left(\int_0^{u_\varepsilon(t,x)} sf'(s)ds\right)dx}_{=0}$$

$$= -\varepsilon \int_{\mathbb{R}} (\partial_x u_\varepsilon)^2\,dx,$$

that is

$$\frac{d}{dt}\|u_\varepsilon(t,\cdot)\|^2_{L^2(\mathbb{R})} + 2\varepsilon \|\partial_x u_\varepsilon(t,\cdot)\|^2_{L(\mathbb{R})} = 0,$$

an integration on $(0, t)$ and (8.25) give (8.27).

We want to prove the existence of converging subsequence using the compensated compactness (see Theorem 8.7). We fix $T > 0$ and an entropy $\eta \in C^2(\mathbb{R})$ with flux q defined by $q' = f'\eta'$. We prove that

$$\{\partial_t \eta(u_\varepsilon) + \partial_x q(u_\varepsilon)\}_{\varepsilon>0} \text{ is compact in } H^{-1}((0,T)\times\mathbb{R}). \tag{8.28}$$

We observe that

$$\partial_t \eta(u_\varepsilon) + \partial_x q(u_\varepsilon) = \eta'(u_\varepsilon)(\partial_t u_\varepsilon + \partial_x f(u_\varepsilon))$$

$$= \varepsilon\eta'(u_\varepsilon)\partial_{xx}^2 u_\varepsilon = \varepsilon\partial_{xx}^2 \eta(u_\varepsilon) - \varepsilon\eta''(u_\varepsilon)(\partial_x u_\varepsilon)^2. \tag{8.29}$$

Since

$$\varepsilon\partial_{xx}^2 \eta(u_\varepsilon) = \partial_x(\varepsilon\eta'(u_\varepsilon)\partial_x u_\varepsilon),$$

and, from (8.27),

$$\left\| \varepsilon \eta'(u_\varepsilon) \partial_x u_\varepsilon \right\|^2_{L^2((0,T) \times \mathbb{R})}$$

$$\leq \left\| \eta' \right\|^2_{L^\infty(-\|u_0\|_{L^\infty(\mathbb{R})}, \|u_0\|_{L^\infty(\mathbb{R})})} \varepsilon^2 \int_0^T \left\| \partial_x u_\varepsilon(s, \cdot) \right\|^2_{L^2(\mathbb{R})} ds$$

$$\leq \left\| \eta' \right\|^2_{L^\infty(-\|u_0\|_{L^\infty(\mathbb{R})}, \|u_0\|_{L^\infty(\mathbb{R})})} \varepsilon \left\| u_0 \right\|^2_{L^2(\mathbb{R})} \longrightarrow 0,$$

therefore

$$\{\varepsilon \partial^2_{xx} \eta(u_\varepsilon)\}_{\varepsilon > 0} \text{ is compact in } H^{-1}((0,T) \times \mathbb{R}). \tag{8.30}$$

Moreover,

$$\left\| \varepsilon \eta''(u_\varepsilon) (\partial_x u_\varepsilon)^2 \right\|_{L^1((0,T) \times \mathbb{R})}$$

$$\leq \left\| \eta'' \right\|_{L^\infty(-\|u_0\|_{L^\infty(\mathbb{R})}, \|u_0\|_{L^\infty(\mathbb{R})})} \varepsilon \int_0^T \left\| \partial_x u_\varepsilon(s, \cdot) \right\|^2_{L^2(\mathbb{R})} ds$$

$$\leq \frac{1}{2} \left\| \eta'' \right\|_{L^\infty(-\|u_0\|_{L^\infty(\mathbb{R})}, \|u_0\|_{L^\infty(\mathbb{R})})} \left\| u_0 \right\|^2_{L^2(\mathbb{R})},$$

therefore

$$\{\varepsilon \eta''(u_\varepsilon)(\partial_x u_\varepsilon)^2\}_{\varepsilon > 0} \text{ is compact in } \mathcal{M}((0,T) \times \mathbb{R}). \tag{8.31}$$

Theorem 8.2 says

$$\{\varepsilon \partial^2_{xx} \eta(u_\varepsilon) + \varepsilon \eta''(u_\varepsilon)(\partial_x u_\varepsilon)^2\}_{\varepsilon > 0} \text{ is compact in } H^{-1}((0,T) \times \mathbb{R}), \tag{8.32}$$

that gives (8.28). Thanks to Theorem 8.7 there exist a subsequence $\{u_{\varepsilon_k}\}_{k \in \mathbb{N}}$ and a function $u \in L^\infty((0, \infty) \times \mathbb{R})$ such that

$$u_{\varepsilon_k} \longrightarrow u \quad \text{in } L^p_{loc}((0, \infty) \times \mathbb{R}), \ 1 \leq p < \infty, \text{ and a.e. in } \Omega. \tag{8.33}$$

Let us show that u is an entropy weak solution of (8.1). Let $\phi \in C^\infty(\mathbb{R}^2)$ be a positive text function with compact support, we have to prove that

$$\int_0^\infty \int_{\mathbb{R}} (\eta(u) \partial_t \phi + q(u) \partial_x \phi) \, dt dx + \int_{\mathbb{R}} \eta(u_0(x)) \phi(0, x) dx \geq 0. \tag{8.34}$$

From (8.29) we have

$$\partial_t \eta(u_{\varepsilon_k}) + \partial_x q(u_{\varepsilon_k}) \leq \varepsilon_k \partial^2_{xx} \eta(u_{\varepsilon_k}).$$

Multiplying by ϕ and integrating on $(0, \infty) \times \mathbb{R}$ we have that

$$\int_0^\infty \int_\mathbb{R} \left(\eta(u_{\varepsilon_k}) \partial_t \phi + q(u_{\varepsilon_k}) \partial_x \phi \right) dt dx$$

$$+ \int_\mathbb{R} \eta(u_{0,\varepsilon_k}(x)) \phi(0, x) dx \qquad (8.35)$$

$$+ \varepsilon_k \int_0^\infty \int_\mathbb{R} \eta(u_{\varepsilon_k}) \partial_{xx}^2 \phi dt dx \geq 0.$$

Therefore, (8.34) follows from (8.25), (8.26), (8.35), and the Dominated Convergence Theorem. □

References

1. Agmon, S.: The L_p approach to the Dirichlet problem. I. Regularity theorems. Ann. Scuola Norm. Sup. Pisa (3) **13**, 405–448 (1959)
2. Ball, J.M.: A version of the fundamental theorem for Young measures. In: PDEs and Continuum Models of Phase Transitions (Nice, 1988). Lecture Notes in Physics, vol. 344, pp. 207–215. Springer, Berlin (1989)
3. Brezis, H.: Functional Analysis, Sobolev Spaces and Partial Differential Equations. Universitext. Springer, New York (2011)
4. Dafermos, C.M.: Hyperbolic Conservation Laws in Continuum Physics. Grundlehren der Mathematischen Wissenschaften [Fundamental Principles of Mathematical Sciences], vol. 325, 4th edn. Springer-Verlag, Berlin (2016)
5. Evans, L.C.: Partial Differential Equations. Graduate Studies in Mathematics, vol. 19, 2nd edn. American Mathematical Society, Providence (2010)
6. Murat, F.: L'injection du cône positif de H^{-1} dans $W^{-1,q}$ est compacte pour tout $q < 2$. J. Math. Pures Appl. (9) **60**(3), 309–322 (1981)
7. Rudin, W.: Real and Complex Analysis, 3rd edn. McGraw-Hill Book Co., New York (1987)
8. Stampacchia, G.: Le problème de Dirichlet pour les équations elliptiques du second ordre à coefficients discontinus. Ann. Inst. Fourier (Grenoble) **15**(fasc. 1), 189–258 (1965)
9. Tartar, L.: Compensated compactness and applications to partial differential equations. In: Nonlinear Analysis and Mechanics: Heriot-Watt Symposium, Vol. IV. Research Notes in Mathematics, vol. 39, pp. 136–212. Pitman, Boston/Mass.-London (1979)
10. Vecchi, I.: A note on entropy compactification for scalar conservation laws. Nonlinear Anal. **15**(7), 693–695 (1990)

Chapter 9
Periodic Solutions

Abstract In this chapter we use compensated compactness in order to study the asymptotic decay of periodic solution.

Keywords Compensated compactness · Asymptotic behavior · Periodic solutions

In this chapter, we consider the Cauchy problem

$$\begin{cases} \partial_t u + \partial_x f(u) = 0, & t > 0, \ x \in \mathbb{R}, \\ u(0, x) = u_0(x), & x \in \mathbb{R}, \end{cases} \tag{9.1}$$

under the assumptions

$$f \text{ genuinely nonlinear,} \qquad u_0 \in L^\infty(\mathbb{R}), \qquad u_0 \text{ is 1-periodic.} \tag{9.2}$$

The aim of the chapter is twofold. First of all we study the existence of periodic entropy solutions to (9.1). Since the Kružkov Theorem holds (see Theorem 3.9), we have the well-posedness of the periodic entropy solutions to (9.1). Then we focus on the asymptotic behavior of these periodic entropy solutions.

9.1   Well-Posedness of Periodic Entropy Solutions

The main result of this section is the following theorem.

Theorem 9.1 *Let f and u_0 satisfy (9.2). The Cauchy problem (9.1) possesses an unique entropy solution $u \in L^\infty((0, \infty) \times \mathbb{R})$ such that $\|u\|_{L^\infty((0,\infty)\times\mathbb{R})} \le \|u_0\|_{L^\infty(\mathbb{R})}$. Moreover, for almost every $t > 0$, $u(t, \cdot)$ is 1-periodic.*

We prove prove existence of a periodic solution to the Cauchy problem (9.1) by analyzing the limiting behavior of the sequence of smooth functions $\{u_\varepsilon\}_{\varepsilon>0}$, where

© The Author(s), under exclusive license to Springer Nature Singapore Pte Ltd. 2024   113
G. M. Coclite, *Scalar Conservation Laws*, SpringerBriefs in Mathematics,
https://doi.org/10.1007/978-981-97-3984-4_9

each function u_ε is the periodic solution of the viscous problem

$$\partial_t u_\varepsilon + \partial_x f(u_\varepsilon) = \varepsilon \partial_{xx}^2 u_\varepsilon, \qquad t > 0, \; x \in \mathbb{R}, \qquad (9.3)$$

endowed with the initial condition

$$u_\varepsilon(0, x) = u_{0,\varepsilon}(x), \qquad x \in \mathbb{R},$$

where we assume that

$u_{0,\varepsilon} \in C^\infty(\mathbb{R})$, $u_{0,\varepsilon}$ is 1-periodic, for every $\varepsilon > 0$,

$\left\| u_{0,\varepsilon} \right\|_{L^2(0,1)} \leq \| u_0 \|_{L^2(0,1)}$, $\left\| u_{0,\varepsilon} \right\|_{L^\infty(\mathbb{R})} \leq \| u_0 \|_{L^\infty(\mathbb{R})}$, for every $\varepsilon > 0$,

$u_{0,\varepsilon} \to u_0$ in $L^2_{loc}(\mathbb{R})$, as $\varepsilon \to 0$.

$$(9.4)$$

Directly from the Maximum Principle for parabolic equations (see Lemma 7.2) we gain:

$$\| u_\varepsilon \|_{L^\infty((0,\infty)\times\mathbb{R})} \leq \| u_0 \|_{L^\infty(\mathbb{R})} . \qquad (9.5)$$

We continue by proving the periodic version of the energy estimate (8.27).

Lemma 9.1 (L^2 Estimate) *The following bounds*

$$\| u_\varepsilon(t, \cdot) \|_{L^2(0,1)}^2 + 2\varepsilon \int_0^t \| \partial_x u_\varepsilon(s, \cdot) \|_{L^2(0,1)}^2 \, ds \leq \| u_0 \|_{L^2(0,1)}^2, \qquad (9.6)$$

hold for any $\varepsilon > 0$, and $t \geq 0$.

Proof We have that

$$\frac{d}{dt} \int_0^1 \frac{u_\varepsilon^2}{2} dx = \int_0^1 u_\varepsilon \partial_t u_\varepsilon dx$$

$$= \varepsilon \int_0^1 u_\varepsilon \partial_{xx}^2 u_\varepsilon dx - \int_0^1 u_\varepsilon f'(u_\varepsilon) \partial_x u_\varepsilon dx$$

$$= -\varepsilon \int_0^1 (\partial_x u_\varepsilon)^2 dx - \underbrace{\int_0^1 \partial_x \left(\int_0^{u_\varepsilon(t,x)} sf'(s) ds \right) dx}_{=0}$$

$$= -\varepsilon \int_0^1 (\partial_x u_\varepsilon)^2 dx,$$

that is

$$\frac{d}{dt} \|u_\varepsilon(t, \cdot)\|_{L^2(0,1)}^2 + 2\varepsilon \|\partial_x u_\varepsilon(t, \cdot)\|_{L^2(0,1)}^2 = 0,$$

an integration on $(0, t)$ and (9.4) give (9.6).   □

Lemma 9.2 *There exists a subsequence $\{u_{\varepsilon_k}\}_{k\in\mathbb{N}}$ of $\{u_\varepsilon\}_{\varepsilon>0}$ and a limit function*

$$u \in L^\infty((0, T) \times \mathbb{R}), \ T > 0, \ \textit{1-periodic in the space variable,} \tag{9.7}$$

such that

$$u_{\varepsilon_k} \to u \ \textit{a.e. and in } L^p_{loc}((0, \infty) \times \mathbb{R}), \ 1 \le p < \infty. \tag{9.8}$$

Proof Let $\eta : \mathbb{R} \to \mathbb{R}$ be any convex C^2 entropy function, and let $q : \mathbb{R} \to \mathbb{R}$ be the corresponding entropy flux defined by $q'(u) = \eta'(u) f'(u)$. By multiplying the first equation in (9.3) with $\eta'(u_\varepsilon)$ and using the chain rule, we get

$$\partial_t \eta(u_\varepsilon) + \partial_x q(u_\varepsilon) = \underbrace{\varepsilon \partial_{xx}^2 \eta(u_\varepsilon)}_{=:\mathcal{L}_\varepsilon^1} \underbrace{-\varepsilon \eta''(u_\varepsilon) (\partial_x u_\varepsilon)^2}_{=:\mathcal{L}_\varepsilon^2}, \tag{9.9}$$

where $\mathcal{L}_\varepsilon^1, \mathcal{L}_\varepsilon^2$ are distributions. We claim that

$$\begin{aligned}\mathcal{L}_\varepsilon^1 &\to 0 \text{ in } H^{-1}((0, \infty) \times (0, 1)),\\ \mathcal{L}_\varepsilon^2 &\text{ is uniformly bounded in } L^1((0, \infty) \times (0, 1)).\end{aligned} \tag{9.10}$$

Indeed, (9.6) implies

$$\|\varepsilon \partial_x \eta(u_\varepsilon)\|_{L^2((0,\infty)\times(0,1))} \le \sqrt{\frac{\varepsilon}{2}} \|\eta'\|_{L^\infty(I)} \|u_0\|_{L^2(0,1)} \to 0, \tag{9.11}$$

$$\left\|\varepsilon \eta''(u_\varepsilon) (\partial_x u_\varepsilon)^2\right\|_{L^1((0,\infty)\times(0,1))} \le \frac{1}{2} \|\eta''\|_{L^\infty(I)} \|u_0\|_{L^2(0,1)}^2, \tag{9.12}$$

where

$$I = \left(-\|u_0\|_{L^\infty(\mathbb{R})}, \|u_0\|_{L^\infty(\mathbb{R})}\right).$$

Hence, (9.10) follows. Therefore, Theorems 8.2 and 8.7 give the existence of a subsequence $\{u_{\varepsilon_k}\}_{k\in\mathbb{N}}$ and a limit function u satisfying (9.7) such that, as $k \to \infty$, (9.8) holds.

Finally, the periodicity of u follows from the periodicity of the viscous approximations and the pointwise convergence stated in (9.8).   □

Proof of Theorem 9.1 Let $\varphi \in C^\infty(\mathbb{R}_+ \times \mathbb{R})$ be a compactly supported test function. Due to (9.3)

$$\int_0^\infty \int_\mathbb{R} \left(u_\varepsilon \partial_t \varphi + f(u_\varepsilon)\partial_x \varphi + \varepsilon u_\varepsilon \partial_{xx}^2 \varphi \right) dx dt + \int_\mathbb{R} u_{0,\varepsilon}(x)\varphi(0,x) dx = 0.$$

Therefore, (9.4) and Lemma 9.2 say that the function u constructed in Lemma 9.2 is a weak solution of (9.1).

Finally, we have to verify that u satisfies the entropy inequalities. Let $\eta \in C^2(\mathbb{R})$ be a convex entropy with flux q defined by $q'(u) = f'(u)\eta'(u)$. The convexity of η and (9.3) yield

$$\partial_t \eta(u_\varepsilon) + \partial_x q(u_\varepsilon) = \varepsilon \partial_{xx}^2 \eta(u_\varepsilon) \underbrace{-\varepsilon \eta''(u_\varepsilon)(\partial_x u_\varepsilon)^2}_{\leq 0} \leq \varepsilon \partial_{xx}^2 \eta(u_\varepsilon).$$

Therefore, the entropy inequalities follow from Lemma 9.2.

The uniqueness of u follows from Theorem 3.9.   □

9.2 Asymptotic Behavior

The main result of this section is the following theorem.

Theorem 9.2 *Let f and u_0 satisfy (9.2). If $u \in L^\infty((0,\infty) \times \mathbb{R})$ is the unique entropy solution of (9.1), then we have*

$$u(t,\cdot) \longrightarrow \int_0^1 u_0 dx, \qquad \text{a.e. and in } L_{loc}^p(\mathbb{R}), \ 1 \leq p < \infty \text{ as } t \to \infty. \quad (9.13)$$

Proof Let u be the periodic entropic solution of (9.1). Following [1] we introduce the functions

$$u_T(t,x) := u(Tt, Tx), \qquad T, t \geq 0, \ x \in \mathbb{R}.$$

Clearly, u_T is $1/T$ periodic in the space variable.

Since u solves (9.1), u_T satisfies

$$\begin{cases} \partial_t u_T + \partial_x f(u_T) = 0, & (t,x) \in (0,\infty) \times \mathbb{R}, \\ u_T(0,x) = u_0(Tx), & x \in \mathbb{R}. \end{cases} \quad (9.14)$$

Due to (9.5) and (9.6)

$$\|u_T(t, \cdot)\|_{L^2(0,1)} \leq \sqrt{\frac{[T] + 1}{T}} \|u_0\|_{L^2(0,1)}, \tag{9.15}$$

$$\|u_T\|_{L^\infty((0,\infty)\times\mathbb{R})} \leq \|u\|_{L^\infty((0,\infty)\times\mathbb{R})}, \tag{9.16}$$

hold for any $T > 0$, and $t \geq 0$, where $[T]$ is the integer part of T.
Indeed

$$\int_0^1 u_T^2(t, x)dx = \int_0^1 u^2(Tt, Tx)dx = \frac{1}{T}\int_0^T u^2(Tt, x)dx$$

$$\leq \frac{1}{T}\int_0^{[T]+1} u^2(Tt, x)dx = \frac{[T] + 1}{T}\int_0^1 u^2(Tt, x)dx$$

$$\leq \frac{[T] + 1}{T}\int_0^1 u_0^2(x)dx.$$

Let $\eta \in C^2(\mathbb{R})$ be a convex entropy with flux q defined by $q'(u) = f'(u)\eta'(u)$.
We claim that

$$\partial_t \eta(u_T) + \partial_x q(u_T) = -\mu_T, \tag{9.17}$$

for some Radon nonnegative measure μ_T on $(0, \infty) \times \mathbb{R}$ such that

$$\mu_T((0, \infty) \times (0, 1)) \leq \frac{1}{2}\|\eta''\|_{L^\infty(I)}\frac{[T] + 1}{T^3}\|u_0\|_{L^2(0,1)}^2, \tag{9.18}$$

for every $T > 0$.
The well posedness argument of the previous section guarantees that

$$u_\varepsilon \to u \text{ a.e. and in } L^p_{loc}((0, \infty) \times \mathbb{R}), \ 1 \leq p < \infty \text{ as } \varepsilon \to 0, \tag{9.19}$$

where u_ε solves (9.3). Defining

$$u_{\varepsilon,T}(t, x) := u_\varepsilon(Tt, Tx),$$

we have

$$u_{\varepsilon,T} \to u_T \text{ a.e. and in } L^p_{loc}((0, \infty) \times \mathbb{R}), \ 1 \leq p < \infty \text{ as } \varepsilon \to 0, \tag{9.20}$$

and $u_{\varepsilon,T}$ solves

$$
\begin{cases}
\partial_t u_{\varepsilon,T} + \partial_x \left(f(u_{\varepsilon,T}) \right) = \frac{\varepsilon}{T} \partial^2_{xx} u_{\varepsilon,T}, & t > 0,\ x \in \mathbb{R}, \\
u_{\varepsilon,T}(0,x) = u_{0,\varepsilon}(Tx), & x \in \mathbb{R}.
\end{cases} \tag{9.21}
$$

We have

$$
\begin{aligned}
\partial_t \eta(u_{\varepsilon,T}) &+ \partial_x q(u_{\varepsilon,T}) \\
&= \frac{\varepsilon}{T} \partial^2_{xx} \eta(u_{\varepsilon,T}) - \frac{\varepsilon}{T} \eta''(u_{\varepsilon,T})(\partial_x u_{\varepsilon,T})^2.
\end{aligned} \tag{9.22}
$$

Since (see (9.6))

$$
\begin{aligned}
\frac{\varepsilon}{T} \int_0^\infty \int_0^1 &\eta''(u_{\varepsilon,T})(\partial_x u_{\varepsilon,T})^2 dt dx \\
&= \frac{\varepsilon}{T^3} \int_0^\infty \int_0^T \eta''(u_\varepsilon)(\partial_x u_\varepsilon)^2 dt dx \\
&\leq \frac{\varepsilon}{T^3} \int_0^\infty \int_0^{[T]+1} \eta''(u_\varepsilon)(\partial_x u_\varepsilon)^2 dt dx \\
&\leq \frac{1}{2} \|\eta''\|_{L^\infty(I)} \frac{[T]+1}{T^3} \|u_0\|^2_{L^2(0,1)},
\end{aligned} \tag{9.23}
$$

as $\varepsilon \to 0$ we get (9.17) and (9.18).

We now use again the argument of the proof of Lemma 9.2 for the family $\{u_T\}_{T>0}$. Thanks to (9.16) and (9.18), we have that $\{\partial_t \eta(u_T) + \partial_x q(u_T)\}_{T>0}$ is bounded in $\mathcal{M}^1_{loc}((0,\infty) \times \mathbb{R})$. Therefore, Theorems 8.2 and 8.7 give the existence of a subsequence $\{u_{T_k}\}_{k \in \mathbb{N}}$, $T_k \to \infty$, and a limit function $u^* \in L^\infty((0,\infty) \times \mathbb{R})$ such that as $k \to \infty$

$$
u_{T_k} \to u^* \text{ a.e. and in } L^p_{loc}((0,\infty) \times \mathbb{R}),\ 1 \leq p < \infty. \tag{9.24}
$$

Therefore, u^* is a weak solution of

$$
\begin{cases}
\partial_t u^* + \partial_x \left(f(u^*) \right) = 0, & (t,x) \in (0,\infty) \times \mathbb{R}, \\
u^*(0,x) = \int_0^1 u_0(x)dx, & x \in \mathbb{R}.
\end{cases} \tag{9.25}
$$

Due to (9.17) and (9.18), we have[1]

$$
\partial_t \eta(u^*) + \partial_x q(u^*) = 0.
$$

---

[1] Here we have the entropy conservation because u^* is smooth.

Then u^* is the entropy solution of (9.25), namely

$$u^* = \int_0^1 u_0(x)dx. \tag{9.26}$$

The uniqueness of u^*, (7.6), and (9.24) guarantee

$$u_T \to u^* \text{ a.e. and in } L^p_{loc}((0, \infty) \times \mathbb{R}), \ 1 \le p < \infty \text{ as } T \to \infty. \tag{9.27}$$

Since

$$\int_0^1 |u_T(t, x) - u^*|dx = \int_0^1 |u(Tt, Tx) - u^*|dx = \frac{1}{T}\int_0^T |u(Tt, x) - u^*|dx,$$

we have

$$\frac{[T]}{T}\int_0^1 |u(Tt, x) - u^*|dx \le \int_0^1 |u_T(t, x) - u^*|dx$$

$$\le \frac{[T] + 1}{T}\int_0^1 |u(Tt, x) - u^*|dx.$$

Therefore, (9.27) implies (9.13).   □

Reference

1. Chen, G.-Q., Frid, H.: Decay of entropy solutions of nonlinear conservation laws. Arch. Ration. Mech. Anal. **146**(2), 95–127 (1999)

Chapter 10
Oleinik Estimate

Abstract This chapter is dedicated to the Oleinik estimate, that is a one side Lipschitz inequality that holds if the flux is strictly convex and is equivalent to the entropy inequalities.

Keywords Oleinik estimate · One side Lipschitz estimate · Uniqueness · Strictly convex fluxes

In this chapter we show a criterion for the uniqueness of weak solutions for the Cauchy problem

$$\begin{cases} \partial_t u + \partial_x f(u) = 0, & t > 0, \ x \in \mathbb{R}, \\ u(0, x) = u_0(x), & x \in \mathbb{R}, \end{cases} \qquad (10.1)$$

when

$$u_0 \in L^1(\mathbb{R}) \cap L^\infty(\mathbb{R}) \qquad \text{and} \qquad f \text{ is strictly convex.}$$

We know that the initial value problem (10.1) admits a unique entropy weak solution (see Theorem 3.9). Moreover, for every $\varepsilon > 0$, we can consider the solution

$$u_\varepsilon \in C^\infty([0, \infty) \times \mathbb{R}) \cap W^{2,p}(0, \infty; W^{1,p}(\mathbb{R})), \qquad 1 \le p < \infty,$$

of the parabolic problem

$$\begin{cases} \partial_t u_\varepsilon + \partial_x f(u_\varepsilon) = \varepsilon \partial_{xx}^2 u_\varepsilon, & t > 0, \ x \in \mathbb{R}, \\ u_\varepsilon(0, x) = u_{0,\varepsilon}(x), & x \in \mathbb{R}, \end{cases} \qquad (10.2)$$

where $u_{0,\varepsilon}$ is a smooth approximation of u_0 such that

$$u_{0,\varepsilon} \in C^\infty(\mathbb{R}) \cap W^{2,1}(\mathbb{R}), \quad \varepsilon > 0,$$

$$u_{0,\varepsilon} \longrightarrow u_0, \quad \text{in } L^p_{loc}(\mathbb{R}), \ 1 \le p < \infty, \ \text{as } \varepsilon \to 0, \tag{10.3}$$

$$\|u_{0,\varepsilon}\|_{L^\infty(\mathbb{R})} \le \|u_0\|_{L^\infty(\mathbb{R})}, \quad \|u_{0,\varepsilon}\|_{L^2(\mathbb{R})} \le \|u_0\|_{L^2(\mathbb{R})}, \quad \varepsilon > 0.$$

A compensated compactness based argument (see Chap. 8) and the Kružkov uniqueness theorem (see Theorem 3.9) say

$$u_\varepsilon \longrightarrow u \quad \text{in } L^p_{loc}((0,\infty) \times \mathbb{R}), \ 1 \le p < \infty, \ \text{and a.e. in } (0,\infty) \times \mathbb{R}$$

$$\text{as } \varepsilon \to 0, \ \text{where } u \text{ is the unique entropy weak solution of (10.1).} \tag{10.4}$$

10.1   Oleinik Estimate

The main result of this section is the following [1].

Theorem 10.1 *Let $u_0 : \mathbb{R} \to \mathbb{R}$ and $f \in C^2(\mathbb{R})$. If*

$$u_0 \in L^1(\mathbb{R}) \cap L^\infty(\mathbb{R}), \qquad f'' \ge c > 0, \quad \text{for some constant } c, \tag{10.5}$$

then

$$\frac{u(t,x) - u(t,y)}{x - y} \le \frac{1}{ct}, \tag{10.6}$$

for almost every $t > 0$ and $x, y \in \mathbb{R}$, $x \ne y$, where u is the unique entropy weak solution of (10.1).

The one side estimate (10.6) is often termed *Oleinik estimate*. We point out that it is coherent with the fact that, due to the convexity of the flux f, we have only downward jumps in the entropy weak solutions of (10.1).

Proof of Theorem 10.1 Let u_ε be the solution of (10.2). We claim that

$$\partial_x u_\varepsilon(t,x) \le \frac{1}{ct}, \qquad t > 0, \ x \in \mathbb{R}. \tag{10.7}$$

Differentiating with respect to x the equation in (10.2) we get

$$\partial^2_{tx} u_\varepsilon + f'(u_\varepsilon)\partial^2_{xx} u_\varepsilon + f''(u_\varepsilon)(\partial_x u_\varepsilon)^2 = \varepsilon \partial^3_{xxx} u_\varepsilon.$$

Let us consider the Cauchy problem

$$\begin{cases} \partial_t v + f'(u_\varepsilon)\partial_x v + f''(u_\varepsilon)v^2 = \varepsilon\partial_{xx}^2 v, & t > 0, \ x \in \mathbb{R}, \\ v(0, x) = u'_{0,\varepsilon}(x), & x \in \mathbb{R}. \end{cases} \qquad (10.8)$$

Clearly the solution of (10.8) is $\partial_x u_\varepsilon$. The map

$$w(t, x) = \frac{1}{ct}, \qquad t > 0, \ x \in \mathbb{R},$$

is a supersolution of (10.8). Indeed, thanks to (10.5),

$$\partial_t w + f'(u_\varepsilon)\partial_x w + f''(u_\varepsilon)w^2 - \varepsilon\partial_{xx}^2 w$$

$$= \partial_t w + f''(u_\varepsilon)w^2$$

$$= -\frac{1}{ct^2} + \frac{f''(u_\varepsilon)}{c^2t^2} = \frac{f''(u_\varepsilon) - c}{c^2t^2} \geq 0,$$

and, at time $t = 0$,

$$\infty = w(0, x) \geq u'_{0,\varepsilon}(x), \qquad x \in \mathbb{R}.$$

Therefore the comparison principle for parabolic equations gives (10.7).
Since for every $t > 0$ and $x, y \in \mathbb{R}$, $x \neq y$, thanks to (10.7),

$$\frac{u_\varepsilon(t, x) - u_\varepsilon(t, y)}{x - y} = \frac{1}{x - y}\int_y^x \partial_x u_\varepsilon(t, \xi)d\xi \leq \frac{1}{ct},$$

(10.4) gives (10.6).   □

10.2 Uniqueness

In this section we show that the Oleinik estimate is equivalent to the entropy conditions and can be used to select the unique entropy weak solution of (10.1) [1, 2].

Theorem 10.2 *Let $u_0 \in L^1(\mathbb{R}) \cap L^\infty(\mathbb{R})$ and $f \in C^2(\mathbb{R})$. If (10.5) holds, there exists at most one distributional solution u of (10.1) such that*

$$u \in L^\infty((0, \infty) \times \mathbb{R}), \qquad \frac{u(t, x) - u(t, y)}{x - y} \leq k\left(\frac{1}{t} + 1\right), \qquad (10.9)$$

for almost every $t > 0$ and $x, y \in \mathbb{R}$, $x \neq y$, and some constant $k > 0$.

Proof Let us assume that there exist two bounded distributional solutions u and v of (10.1) such that

$$\frac{u(t, x) - u(t, y)}{x - y} \leq k \left(\frac{1}{t} + 1\right), \qquad \frac{v(t, x) - v(t, y)}{x - y} \leq k \left(\frac{1}{t} + 1\right),$$

$$(10.10)$$

for almost every $t > 0$ and $x, y \in \mathbb{R}$, $x \neq y$, and some constant $k > 0$.

We claim that

$$u = v, \qquad \text{a.e. in } [0, \infty) \times \mathbb{R}. \qquad (10.11)$$

Let $\varphi \in C^\infty(\mathbb{R}^2)$ be a test function with compact support. Since u and v are distributional solutions of (10.1) we have that

$$\int_0^\infty \int_{\mathbb{R}} (u \partial_t \varphi + f(u) \partial_x \varphi) \, dt dx + \int_{\mathbb{R}} u_0(x) \varphi(0, x) dx = 0,$$

$$\int_0^\infty \int_{\mathbb{R}} (v \partial_t \varphi + f(v) \partial_x \varphi) \, dt dx + \int_{\mathbb{R}} u_0(x) \varphi(0, x) dx = 0,$$

and then

$$0 = \int_0^\infty \int_{\mathbb{R}} \Big((u - v) \partial_t \varphi + (f(u) - f(v)) \partial_x \varphi \Big) dt dx$$

$$= \int_0^\infty \int_{\mathbb{R}} w \, (\partial_t \varphi + b \partial_x \varphi) \, dt dx,$$

$$(10.12)$$

where

$$w = u - v,$$

$$b(t, x) = \int_0^1 f'(\theta u(t, x) + (1 - \theta)v(t, x)) d\theta = \frac{f(u(t, x)) - f(v(t, x))}{u(t, x) - v(t, x)}.$$

The *idea of the proof* is the following. Let $\psi \in C^\infty(\mathbb{R}^2)$ be a test function with compact support and let $\varphi \in C^\infty(\mathbb{R}^2)$ be a solution of the linear transport equation

$$\partial_t \varphi + b \partial_x \varphi = \psi. \qquad (10.13)$$

From (10.12), we get

$$\int_0^\infty \int_{\mathbb{R}} (u - v) \psi \, dt dx = \int_0^\infty \int_{\mathbb{R}} w \, (\partial_t \varphi + b \partial_x \varphi) \, dt dx = 0,$$

and then $u = v$. Unfortunately we do not have enough regularity on b therefore we are not able to solve (10.13). We have to use a sharper argument [1, 2].

We begin by regularizing b and w. Let $\{\rho_\varepsilon\}_{\varepsilon>0} \subset C^\infty(\mathbb{R}^2)$ a family of mollifiers in t and x such that

$$\text{supp}(\rho_\varepsilon) \subset (-\infty, 0) \times \mathbb{R}, \qquad \varepsilon > 0.$$

Let $\varepsilon > 0$. We write

$$u_\varepsilon := \rho_\varepsilon * u, \qquad v_\varepsilon := \rho_\varepsilon * v, \qquad w_\varepsilon := \rho_\varepsilon * w,$$

$$b_\varepsilon(t, x) := \int_0^1 f'(\theta u_\varepsilon(t, x) + (1 - \theta)v_\varepsilon(t, x))d\theta,$$

where $*$ denotes the convolution in both variables t and x. Clearly

$w_\varepsilon = u_\varepsilon - v_\varepsilon,$

$\|u_\varepsilon\|_{L^\infty((0,\infty)\times\mathbb{R})} \leq \|u\|_{L^\infty((0,\infty)\times\mathbb{R})}, \qquad \|v_\varepsilon\|_{L^\infty((0,\infty)\times\mathbb{R})} \leq \|v\|_{L^\infty((0,\infty)\times\mathbb{R})},$

$u_\varepsilon \longrightarrow u,\ v_\varepsilon \longrightarrow v, \quad$ a.e. in $(0, \infty) \times \mathbb{R}$ and in $L^1((0, \infty) \times \mathbb{R})$,

$w_\varepsilon \longrightarrow w,\ b_\varepsilon \longrightarrow b, \quad$ a.e. in $(0, \infty) \times \mathbb{R}$ and in $L^1((0, \infty) \times \mathbb{R})$.

Moreover, we have that

$$\partial_x u_\varepsilon(t, x) \leq k\left(\frac{1}{t} + 1\right), \qquad \partial_x v_\varepsilon(t, x) \leq k\left(\frac{1}{t} + 1\right), \qquad t > 0,\ x \in \mathbb{R}. \tag{10.14}$$

Indeed

$$\frac{u_\varepsilon(t, x) - u_\varepsilon(t, y)}{x - y} = \int_{-\infty}^0 \int_{\mathbb{R}} \rho_\varepsilon(s, \xi) \frac{u(t - s, x - \xi) - u(t - s, y - \xi)}{x - y} ds d\xi$$

$$\leq k\left(\int_{-\infty}^0 \int_{\mathbb{R}} \frac{\rho_\varepsilon(s, \xi)}{t - s} ds d\xi + \int_{-\infty}^0 \int_{\mathbb{R}} \rho_\varepsilon(s, \xi) ds d\xi\right)$$

$$\leq k\left(\frac{1}{t} + 1\right) \int_{-\infty}^0 \int_{\mathbb{R}} \rho_\varepsilon(s, \xi) ds d\xi = k\left(\frac{1}{t} + 1\right),$$

as $y \to x$ we get (10.14).

Since

$\partial_x b_\varepsilon(t, x)$

$$= \int_0^1 f''(\theta u_\varepsilon(t, x) + (1 - \theta)v_\varepsilon(t, x)) \cdot (\theta \partial_x u_\varepsilon(t, x) + (1 - \theta)\partial_x v_\varepsilon(t, x))d\theta,$$

(10.14), the convexity and smoothness of f, and the boundedness of u_ε and v_ε give

$$\partial_x b_\varepsilon(t, x) \leq k' \left(\frac{1}{t} + 1 \right), \qquad t > 0, \ x \in \mathbb{R}, \tag{10.15}$$

where

$$k' = k \| f'' \|_{L^\infty(-M, M)}, \qquad M = \| u \|_{L^\infty((0,\infty) \times \mathbb{R})} + \| v \|_{L^\infty((0,\infty) \times \mathbb{R})}.$$

Let $\psi \in C^\infty((0, \infty) \times \mathbb{R})$ be a test function with compact support and $T > \tau > 0$ be such that

$$\text{supp}(\psi) \subset [\tau, T] \times \mathbb{R}, \tag{10.16}$$

and, let φ_ε be the solution of the backward transport problem

$$\begin{cases} \partial_t \varphi_\varepsilon + b_\varepsilon \partial_x \varphi_\varepsilon = \psi, & 0 < t < T, \ x \in \mathbb{R}, \\ \varphi_\varepsilon(T, x) = 0, & x \in \mathbb{R}. \end{cases} \tag{10.17}$$

In light of the regularity of b_ε and ψ, using the method of the characteristics, we have that

$$\varphi_\varepsilon \in C^\infty((0, T) \times \mathbb{R}) \text{ with compact support.} \tag{10.18}$$

Thanks to (10.12) and (10.17)

$$\int_0^T \int_\mathbb{R} w\psi \, dt dx = \int_0^T \int_\mathbb{R} w(\partial_t \varphi_\varepsilon + b_\varepsilon \partial_x \varphi_\varepsilon) dt dx$$

$$= \underbrace{\int_0^T \int_\mathbb{R} w(\partial_t \varphi_\varepsilon + b \partial_x \varphi_\varepsilon) dt dx}_{=0}$$

$$+ \int_0^T \int_\mathbb{R} w(b_\varepsilon - b) \partial_x \varphi_\varepsilon \, dt dx \tag{10.19}$$

$$= \int_0^T \int_\mathbb{R} w(b_\varepsilon - b) \partial_x \varphi_\varepsilon \, dt dx.$$

Therefore, we have to prove that

$$\lim_{\varepsilon \to 0} \int_0^T \int_\mathbb{R} w(b_\varepsilon - b) \partial_x \varphi_\varepsilon \, dt dx = 0. \tag{10.20}$$

We begin by proving some a priori estimates on φ_ε. Due to regularity of φ_ε, b_ε, ψ we can differentiate the equation in (10.17) with respect to x

$$\partial^2_{tx}\varphi_\varepsilon + \partial_x b_\varepsilon \partial_x \varphi_\varepsilon + b_\varepsilon \partial^2_{xx}\varphi_\varepsilon = \partial_x \psi, \tag{10.21}$$

therefore

$$\frac{d}{dt}\int_{\mathbb{R}}|\partial_x\varphi_\varepsilon|dx = \int_{\mathbb{R}} \partial^2_{tx}\varphi_\varepsilon \operatorname{sign}(\partial_x\varphi_\varepsilon)\,dx$$

$$= \int_{\mathbb{R}} \partial_x\psi \operatorname{sign}(\partial_x\varphi_\varepsilon)\,dx$$

$$- \int_{\mathbb{R}} \partial_x b_\varepsilon \underbrace{\partial_x\varphi_\varepsilon \operatorname{sign}(\partial_x\varphi_\varepsilon)}_{=|\partial_x\varphi_\varepsilon|}\,dx - \int_{\mathbb{R}} b_\varepsilon \underbrace{\partial^2_{xx}\varphi_\varepsilon \operatorname{sign}(\partial_x\varphi_\varepsilon)}_{=\partial_x|\partial_x\varphi_\varepsilon|}\,dx$$

$$= \int_{\mathbb{R}} \partial_x\psi \operatorname{sign}(\partial_x\varphi_\varepsilon)\,dx - \int_{\mathbb{R}} \partial_x b_\varepsilon |\partial_x\varphi_\varepsilon|dx - \int_{\mathbb{R}} b_\varepsilon \partial_x|\partial_x\varphi_\varepsilon|dx$$

$$= \int_{\mathbb{R}} \partial_x\psi \operatorname{sign}(\partial_x\varphi_\varepsilon)\,dx$$

$$\geq - \int_{\mathbb{R}} |\partial_x\psi|dx,$$

and, integrating on $[t, T]$,

$$\|\partial_x\varphi_\varepsilon(t, \cdot)\|_{L^1(\mathbb{R})} \leq \|\partial_x\psi\|_{L^1((0,T)\times\mathbb{R})}, \qquad 0 \leq t \leq T. \tag{10.22}$$

We prove an L^∞ estimate on $\partial_x\varphi_\varepsilon$. Let $p \in \mathbb{N} \setminus \{0\}$ be even. Thanks to (10.15), (10.21), and the Hölder inequality,

$$\frac{d}{dt}\|\partial_x\varphi_\varepsilon(t, \cdot)\|^p_{L^p(\mathbb{R})} = \frac{d}{dt}\int_{\mathbb{R}}(\partial_x\varphi_\varepsilon)^p dx = p\int_{\mathbb{R}}(\partial_x\varphi_\varepsilon)^{p-1}\partial^2_{tx}\varphi_\varepsilon dx$$

$$= p\int_{\mathbb{R}} \partial_x\psi (\partial_x\varphi_\varepsilon)^{p-1}dx - p\int_{\mathbb{R}} \partial_x b_\varepsilon (\partial_x\varphi_\varepsilon)^p dx$$

$$- p\int_{\mathbb{R}} b_\varepsilon \underbrace{\partial^2_{xx}\varphi_\varepsilon(\partial_x\varphi_\varepsilon)^{p-1}}_{=\partial_x((\partial_x\varphi_\varepsilon)^p)/p}\,dx$$

$$= p\int_{\mathbb{R}} \partial_x\psi (\partial_x\varphi_\varepsilon)^{p-1}dx - p\int_{\mathbb{R}} \partial_x b_\varepsilon (\partial_x\varphi_\varepsilon)^p dx - \int_{\mathbb{R}} b_\varepsilon \partial_x((\partial_x\varphi_\varepsilon)^p)dx$$

$$= p\int_{\mathbb{R}} \partial_x\psi (\partial_x\varphi_\varepsilon)^{p-1}dx - (p-1)\int_{\mathbb{R}} \partial_x b_\varepsilon (\partial_x\varphi_\varepsilon)^p dx$$

$$\geq -p\|\partial_x\psi(t, \cdot)\|_{L^p(\mathbb{R})} \left\|(\partial_x\varphi_\varepsilon(t, \cdot))^{p-1}\right\|_{L^{\frac{p}{p-1}}(\mathbb{R})}$$

$$- (p-1)k' \left(\frac{1}{t} + 1 \right) \| \partial_x \varphi_\varepsilon (t, \cdot) \|_{L^p(\mathbb{R})}^p$$

$$\geq -p\alpha \, \| \partial_x \varphi_\varepsilon (t, \cdot)) \|_{L^p(\mathbb{R})}^{p-1} - (p-1)k' \left(\frac{1}{t} + 1 \right) \| \partial_x \varphi_\varepsilon (t, \cdot) \|_{L^p(\mathbb{R})}^p,$$

for some constant $\alpha > 0$ independent on ε. Therefore we have

$$p \, \| \partial_x \varphi_\varepsilon (t, \cdot) \|_{L^p(\mathbb{R})}^{p-1} \frac{d}{dt} \, \| \partial_x \varphi_\varepsilon (t, \cdot) \|_{L^p(\mathbb{R})}$$

$$\geq -p\alpha \, \| \partial_x \varphi_\varepsilon (t, \cdot) \|_{L^p(\mathbb{R})}^{p-1} - (p-1)k' \left(\frac{1}{t} + 1 \right) \| \partial_x \varphi_\varepsilon (t, \cdot) \|_{L^p(\mathbb{R})}^p,$$

that is

$$\frac{d}{dt} \, \| \partial_x \varphi_\varepsilon (t, \cdot) \|_{L^p(\mathbb{R})} \geq -\alpha - \frac{p-1}{p} k' \left(\frac{1}{t} + 1 \right) \| \partial_x \varphi_\varepsilon (t, \cdot) \|_{L^p(\mathbb{R})}.$$

The Gronwall Lemma and the final condition in (10.17) give

$$0 = \| \partial_x \varphi_\varepsilon (T, \cdot) \|_{L^p(\mathbb{R})}$$

$$\geq \| \partial_x \varphi_\varepsilon (t, \cdot) \|_{L^p(\mathbb{R})} \, e^{-k' \frac{p-1}{p} \left(\log \left(\frac{T}{t} \right) + T - t \right)}$$

$$- \alpha e^{-k' \frac{p-1}{p} \left(\log \left(\frac{T}{t} \right) + T - t \right)} \int_t^T e^{k' \frac{p-1}{p} \left(\log \left(\frac{s}{t} \right) + s - t \right)} ds$$

$$= \| \partial_x \varphi_\varepsilon (t, \cdot) \|_{L^p(\mathbb{R})} \left(\frac{t}{T} \right)^{k' \frac{p-1}{p}} e^{-k' \frac{p-1}{p} (T-t)}$$

$$- \alpha \left(\frac{t}{T} \right)^{k' \frac{p-1}{p}} e^{-k' \frac{p-1}{p} (T-t)} \int_t^T \left(\frac{s}{t} \right)^{k' \frac{p-1}{p}} e^{k' \frac{p-1}{p} (s-t)} ds$$

$$= \| \partial_x \varphi_\varepsilon (t, \cdot) \|_{L^p(\mathbb{R})} \left(\frac{t}{T} \right)^{k' \frac{p-1}{p}} e^{-k' \frac{p-1}{p} (T-t)} - \alpha \int_t^T \left(\frac{s}{T} \right)^{k' \frac{p-1}{p}} e^{-k' \frac{p-1}{p} (T-s)} ds$$

$$\geq \| \partial_x \varphi_\varepsilon (t, \cdot) \|_{L^p(\mathbb{R})} \left(\frac{t}{T} \right)^{k' \frac{p-1}{p}} e^{-k' \frac{p-1}{p} (T-t)} - \alpha (T-t)$$

$$\geq \| \partial_x \varphi_\varepsilon (t, \cdot) \|_{L^p(\mathbb{R})} \left(\frac{t}{T} \right)^{k' \frac{p-1}{p}} e^{-k' \frac{p-1}{p} (T-t)} - \alpha T,$$

that is

$$\| \partial_x \varphi_\varepsilon (t, \cdot) \|_{L^p(\mathbb{R})} \leq \alpha T \left(\frac{T}{t} \right)^{k' \frac{p-1}{p}} e^{k' \frac{p-1}{p} (T-t)}.$$

As $p \to \infty$

$$\|\partial_x \varphi_\varepsilon(t, \cdot)\|_{L^\infty(\mathbb{R})} \le \alpha \frac{T^{k'+1}}{t^{k'}} e^{k'(T-t)}. \tag{10.23}$$

Therefore, for every $0 < s < T$, using (10.22) and (10.23)

$$\left| \int_0^T \int_{\mathbb{R}} w(b - b_\varepsilon) \partial_x \varphi_\varepsilon \, dt dx \right|$$

$$\le \int_0^s \int_{\mathbb{R}} |w||b - b_\varepsilon||\partial_x \varphi_\varepsilon| dt dx + \int_s^T \int_{\mathbb{R}} |w||b - b_\varepsilon||\partial_x \varphi_\varepsilon| dt dx$$

$$\le \|w\|_{L^\infty((0,T) \times \mathbb{R})} \left(\|b\|_{L^\infty((0,T) \times \mathbb{R})} + \|b_\varepsilon\|_{L^\infty((0,T) \times \mathbb{R})} \right) \|\partial_x \varphi_\varepsilon\|_{L^1((0,s) \times \mathbb{R})}$$

$$+ \|w\|_{L^\infty((0,T) \times \mathbb{R})} \|b - b_\varepsilon\|_{L^1((0,T) \times \mathbb{R})} \|\partial_x \varphi_\varepsilon\|_{L^\infty((s,T) \times \mathbb{R})}$$

$$\le C \left(s + \frac{\|b - b_\varepsilon\|_{L^1((0,T) \times \mathbb{R})}}{s^{k'} e^{k's}} \right),$$

for some constant $C > 0$ independent on s and ε. As $\varepsilon \to 0$

$$\limsup_{\varepsilon \to 0} \left| \int_0^T \int_{\mathbb{R}} w(b - b_\varepsilon) \partial_x \varphi_\varepsilon \, dt dx \right| \le C s,$$

since the left hand side does not depend on s we can send $s \to 0$ and get (10.20). $\qquad \square$

We conclude this section with the following corollary.

Corollary 10.1 *Let $u_0 : \mathbb{R} \to \mathbb{R}$, $f \in C^2(\mathbb{R})$, and $u : [0, \infty) \times \mathbb{R} \to \mathbb{R}$. If (10.5) holds and u is a distributional solution of (10.1), the following statements are equaivalent*

(i) u is the entropy weak solution of (10.1);
(ii) for almost every $t > 0$ and $x, y \in \mathbb{R}$, $x \ne y$

$$\frac{u(t, x) - u(t, y)}{x - y} \le \frac{1}{ct},$$

where c is the constant introduced in (10.5).

Proof The implication (i) $\Rightarrow$ (ii) has been proved in Theorem 10.1. Let us prove (ii) $\Rightarrow$ (i). We have only to observe that the entropy weak solution of (10.1) satisfies (ii) and that the distributional solution of (10.1) satisfying (ii) is unique. Therefore, the unique distributional solution satisfying (ii) has to be the unique entropy weak solution of (10.1). $\qquad \square$

10.3   Exercises

Exercise 10.1 Let u be the solution of the Cauchy problem

$$\begin{cases} \partial_t u + \partial_x f(u) = \partial_{xx}^2 u + \arctan(x), & t > 0, \ x \in \mathbb{R}, \\ u(0, x) = \frac{1}{1+x^2}, & x \in \mathbb{R}. \end{cases}$$

Assume

$$f \in C^2(\mathbb{R}), \qquad f'' \geq c > 0,$$

prove

$$\partial_x u(t, x) \leq \frac{1}{ct} + \frac{1}{\sqrt{c}}, \qquad t > 0, \ x \in \mathbb{R}.$$

Exercise 10.2 Let u be the solution of the Cauchy problem

$$\begin{cases} \partial_t u + \partial_x f(u) = \partial_{xx}^2 u + e^x, & t > 0, \ x \in \mathbb{R}, \\ u(0, x) = \frac{1}{1+x^2}, & x \in \mathbb{R}. \end{cases}$$

Assume

$$f \in C^2(\mathbb{R}), \qquad f'' \leq -c < 0,$$

prove

$$\partial_x u(t, x) \geq -\frac{1}{ct}, \qquad t > 0, \ x \in \mathbb{R}.$$

References

1. Oleinik, O.A.: Discontinuous solutions of non-linear differential equations. Am. Math. Soc. Transl. (2) **26**, 95–172 (1963)
2. Tadmor, E.: Local error estimates for discontinuous solutions of nonlinear hyperbolic equations. SIAM J. Numer. Anal. **28**(4), 891–906 (1991)

Chapter 11
Lax-Oleinik Formula

Abstract In this chapter we present the Lax-Oleinik Formula, that provides and an explicit formula for the solutions of the initial value problem when the flux is strictly convex.

Keywords Legendre transform · Lax-Oleinik formula · Hamilton-Jacobi equation · Strictly convex fluxes · Representation formula

In this chapter we prove a representation formula for the entropy solutions for the Cauchy problem

$$\begin{cases} \partial_t u + \partial_x f(u) = 0, & t > 0, \ x \in \mathbb{R}, \\ u(0, x) = u_0(x), & x \in \mathbb{R}, \end{cases} \tag{11.1}$$

when

$$u_0 \in L^\infty(\mathbb{R}), \qquad f \in C^3(\mathbb{R}), \qquad f(0) = 0,$$

$$f \text{ is strictly convex}, \qquad \lim_{|u| \to \infty} \frac{f(u)}{|u|} = \infty. \tag{11.2}$$

We proved in Theorem 3.9 that the initial value problem (11.1) admits a unique entropy weak solution. We look for a representation of u in terms of f and u_0. The main tool of the argument is the equivalence between the conservation law

$$\partial_t u + \partial_x f(u) = 0 \tag{11.3}$$

and the Hamilton-Jacobi equation

$$\partial_t v + f(\partial_x v) = 0. \tag{11.4}$$

Indeed, if v is a solution of (11.4), then, differentiating both sides of (11.4), $\partial_x v$ is a solution of (11.3), see [2].

11.1   Legendre Transform

Let $f : \mathbb{R} \to \mathbb{R}$ be a function such that

$$f \text{ is convex} \quad \text{and} \quad \lim_{|u| \to \infty} \frac{f(u)}{|u|} = \infty. \tag{11.5}$$

The *Legendre transform* of f is the map

$$f^* : \mathbb{R} \longrightarrow \mathbb{R}, \qquad f^*(\xi) = \sup_{u \in \mathbb{R}} (u\xi - f(u)).$$

Let $\xi \in \mathbb{R}$. Since the map $u \mapsto u\xi - f(u)$ is continuous and

$$\lim_{|u| \to \infty} (u\xi - f(u)) = \lim_{|u| \to \infty} |u| \left(\text{sign}(u)\,\xi - \frac{f(u)}{u} \right) = -\infty,$$

the definition of f^* makes sense, indeed there exists at least one point $u(\xi) \in \mathbb{R}$ such that

$$f^*(\xi) = u(\xi)\xi - f(u(\xi)). \tag{11.6}$$

If f is differentiable, we have

$$\partial_u (u\xi - f(u)) \Big|_{u=u(\xi)} = 0,$$

therefore

$$\xi = f'(u(\xi)). \tag{11.7}$$

If f is strictly convex, f' is increasing and, in particular, invertible. Writing

$$g = (f')^{-1}, \tag{11.8}$$

(11.7) gives

$$u(\xi) = g(\xi), \tag{11.9}$$

and (11.6) reads

$$f^*(\xi) = g(\xi)\xi - f(g(\xi)). \tag{11.10}$$

Finally, differentiating (11.6) with respect to ξ we have

$$(f^*)'(\xi) = u(\xi) + \underbrace{(\xi - f'(u(\xi)))}_{=0 \text{ (see (11.7))}} \partial_\xi u(\xi),$$

then, thanks to (11.9),

$$(f^*)'(\xi) = u(\xi) = g(\xi). \tag{11.11}$$

Theorem 11.1 *If f satisfies (11.5), then also f^* does, and*

$$(f^*)^* = f. \tag{11.12}$$

Proof Let $\lambda \in [0, 1]$ and ξ, $\xi' \in \mathbb{R}$ be given. We have

$$f^*(\lambda\xi + (1 - \lambda)\xi') = \sup_{u \in \mathbb{R}} \left(u(\lambda\xi + (1 - \lambda)\xi') - f(u) \right)$$

$$\leq \lambda \sup_{u \in \mathbb{R}} (u\xi - f(u)) + (1 - \lambda) \sup_{u \in \mathbb{R}} \left(u\xi' - f(u) \right)$$

$$= \lambda f^*(\xi) + (1 - \lambda) f^*(\xi'),$$

then f^* is convex.

Given $\lambda > 0$ and $\xi \in \mathbb{R} \setminus \{0\}$, since

$$f^*(\xi) = \sup_{u \in \mathbb{R}} (u\xi - f(u)) \geq (u\xi - f(u)) \Big|_{u = \lambda \frac{|\xi|}{\xi}}$$

$$= \lambda|\xi| - f\left(\lambda \frac{|\xi|}{\xi}\right) \geq \lambda|\xi| - \max_{[-\lambda, \lambda]} f,$$

we have

$$\frac{f^*(\xi)}{|\xi|} \geq \lambda - \frac{\max\limits_{[-\lambda, \lambda]} f}{|\xi|}.$$

As $|\xi| \to \infty$, we have

$$\liminf_{|\xi| \to \infty} \frac{f^*(\xi)}{|\xi|} \geq \lambda.$$

Since this holds for every $\lambda > 0$,

$$\lim_{|\xi| \to \infty} \frac{f^*(\xi)}{|\xi|} = \infty.$$

As a consequence f^* satisfies (11.5).

We have to prove (11.12). From the definition of f^* we have

$$f(u) \geq u\xi - f^*(\xi), \qquad u, \xi \in \mathbb{R},$$

then, for every $u \in \mathbb{R}$,

$$f(u) \geq \sup_{\xi \in \mathbb{R}} \left(u\xi - f^*(\xi) \right) = (f^*)^*(u),$$

namely

$$f \geq (f^*)^*. \tag{11.13}$$

Since f is convex, for every $u \in \mathbb{R}$ there exists $c_u \in \mathbb{R}$ such that

$$f(v) \geq f(u) + c_u(v - u), \qquad v \in \mathbb{R}.$$

We have

$$(f^*)^*(u) = \sup_{\xi \in \mathbb{R}} \left(u\xi - f^*(\xi) \right)$$

$$= \sup_{\xi \in \mathbb{R}} \left(u\xi - \sup_{v \in \mathbb{R}} (v\xi - f(v)) \right)$$

$$= \sup_{\xi \in \mathbb{R}} \inf_{v \in \mathbb{R}} (\xi(u - v) + f(v))$$

$$\geq \sup_{\xi \in \mathbb{R}} \inf_{v \in \mathbb{R}} (\xi(u - v) + f(u) + c_u(v - u))$$

$$\geq \inf_{v \in \mathbb{R}} (\xi(u - v) + f(u) + c_u(v - u)) \Big|_{\xi = c_u} = \inf_{v \in \mathbb{R}} f(u) = f(u),$$

that is

$$f \leq (f^*)^*. \tag{11.14}$$

Clearly, (11.13) and (11.14) give (11.12).   □

Theorem 11.2 *Let f satisfy* (11.5). *If*

$$f \in C^3(\mathbb{R}), \qquad f(0) = 0, \qquad f \text{ is strictly convex,} \tag{11.15}$$

then

$$f^* \in C^2(\mathbb{R}), \qquad f^* \geq 0, \qquad f^* \text{ is strictly convex.} \tag{11.16}$$

Proof The regularity of f^* follows from the one of f and (11.10). In addition, for every $\xi \in \mathbb{R}$

$$f^*(\xi) = \sup_{u \in \mathbb{R}} (u\xi - f(u)) \geq (u\xi - f(u)) \Big|_{u=0} = 0.$$

Finally, since f' is strictly increasing, the same holds for g, then from (11.11)

$$(f^*)'' = g' > 0,$$

namely f^* is strictly convex.   □

11.2   Lax–Oleinik Formula

The main result of this section is the following [3, 4].

Theorem 11.3 (Lax–Oleinik Formula) *Let us assume* (11.2) *The following statements hold.*

(i) *For every $t > 0$, there exists for all but at most countably many values of $x \in \mathbb{R}$ a unique $y(t, x)$ such that*

$$\min_{y \in \mathbb{R}} \left(tf^* \left(\frac{x - y}{t} \right) + \int_0^y u_0(s)ds \right) = tf^* \left(\frac{x - y(t, x)}{t} \right)$$

$$+ \int_0^{y(t,x)} u_0(s)ds. \tag{11.17}$$

(ii) *The function $x \mapsto y(t, x)$ is nondecreasing.*
(iii) *For each $t > 0$ and almost every $x \in \mathbb{R}$ the function*

$$u(t, x) = g \left(\frac{x - y(t, x)}{t} \right) \tag{11.18}$$

provides the unique entropy solution of (11.1).

Proof For the sake of notational simplicity we introduce the notation

$$F(t, x, y) = t f^* \left(\frac{x - y}{t} \right) + \int_0^y u_0(s) ds.$$

Let $t > 0$. For every $x \in \mathbb{R}$, since the function

$$y \longmapsto F(t, x, y)$$

is continuous and

$$\lim_{|y| \to \infty} F(t, x, y) = \lim_{|y| \to \infty} |y| \left(\frac{t}{|y|} f^* \left(\frac{x - y}{t} \right) + \frac{1}{|y|} \int_0^y u_0(s) ds \right) = \infty,$$

there exists at least a value $\overline{y} \in \mathbb{R}$ such that

$$F(t, x, \overline{y}) = \min_{y \in \mathbb{R}} F(t, x, y).$$

Moreover, introducing the function

$$y(t, x) = \inf \left\{ \overline{y} \in \mathbb{R}; \ F(t, x, \overline{y}) = \min_{y \in \mathbb{R}} F(t, x, y) \right\},$$

we have

$$y(t, x) \in \mathbb{R},$$

$$y(t, x) = \min \left\{ \overline{y} \in \mathbb{R}; \ F(t, x, \overline{y}) = \min_{y \in \mathbb{R}} F(t, x, y) \right\}.$$

Given $x_1 < x_2$ and y_1 such that

$$F(t, x_1, y_1) = \min_{y \in \mathbb{R}} F(t, x_1, y).$$

We claim that

$$F(t, x_2, y_1) < F(t, x_2, y), \qquad y < y_1. \tag{11.19}$$

Fix $y < y_1$ and define

$$\lambda = \frac{y_1 - y}{x_2 - x_1 + y_1 - y}.$$

We have

$$0 < \lambda < 1,$$
$$x_2 - y_1 = \lambda(x_1 - y_1) + (1 - \lambda)(x_2 - y),$$
$$x_1 - y = \lambda(x_2 - y) + (1 - \lambda)(x_1 - y_1).$$

Due to the strict convexity of f^* we have

$$f^*\left(\frac{x_2 - y_1}{t}\right) < \lambda f^*\left(\frac{x_1 - y_1}{t}\right) + (1 - \lambda)f^*\left(\frac{x_2 - y}{t}\right),$$

$$f^*\left(\frac{x_1 - y}{t}\right) < \lambda f^*\left(\frac{x_2 - y}{t}\right) + (1 - \lambda)f^*\left(\frac{x_1 - y_1}{t}\right),$$

and then

$$f^*\left(\frac{x_2 - y_1}{t}\right) + f^*\left(\frac{x_1 - y}{t}\right) < f^*\left(\frac{x_1 - y_1}{t}\right) + f^*\left(\frac{x_2 - y}{t}\right).$$

Hence

$$F(t, x_2, y_1) = tf^*\left(\frac{x_2 - y_1}{t}\right) + \int_0^{y_1} u_0(s)ds$$

$$< tf^*\left(\frac{x_1 - y_1}{t}\right) + \int_0^{y_1} u_0(s)ds + tf^*\left(\frac{x_2 - y}{t}\right) - tf^*\left(\frac{x_1 - y}{t}\right)$$

$$\leq tf^*\left(\frac{x_1 - y}{t}\right) + \int_0^{y} u_0(s)ds + tf^*\left(\frac{x_2 - y}{t}\right) - tf^*\left(\frac{x_1 - y}{t}\right)$$

$$= tf^*\left(\frac{x_2 - y}{t}\right) + \int_0^{y} u_0(s)ds = F(t, x_2, y),$$

that is (11.19).

From (11.19) we gain

$$\begin{array}{c} \text{for every } x_1 < x_2, \ y_1, \ y_2 \text{ such that} \\[4pt] F(t, x_1, y_1) = \min_{y \in \mathbb{R}} F(t, x_1, y), \\[4pt] F(t, x_2, y_2) = \min_{y \in \mathbb{R}} F(t, x_2, y), \\[4pt] \text{we have } y_1 < y_2. \end{array} \qquad (11.20)$$

As a consequence, the function $x \mapsto y(t, x)$ is nondecreasing, and in particular is continuous in all but at most countably many $x \in \mathbb{R}$.

We claim that

$$\text{if } y(t, \cdot) \text{ is continuous in } x, \text{ then}$$
$$y(t, x) \text{ is the unique minimum point of } F(t, x, \cdot).$$
(11.21)

Let $\overline{y}$ be a minimizer of $F(t, x, \cdot)$ in correspondence of (t, x), due to (11.20),

$$y(t, x) \leq \overline{y} < y\left(t, x + \frac{1}{n}\right), \qquad n \in \mathbb{N} \setminus \{0\}.$$

Due to the continuity of $y(t, \cdot)$ in x we must have $y(t, x) = \overline{y}$.

Since the function $y(t, \cdot)$ is nondecreasing, it is differentiable at almost every $x \in \mathbb{R}$. As a consequence, also the function

$$w(t, x) = F(t, x, y(t, x))$$

is differentiable almost everywhere and

$$\partial_x w(t, x) = (f^*)'\left(\frac{x - y(t, x)}{t}\right)(1 - \partial_x y(t, x)) + u_0(y(t, x))\partial_x y(t, x).$$
(11.22)

On the other hand, since the function

$$y \longmapsto F(t, x, y)$$

has a minimum in $y(t, x)$, then the function

$$z \longmapsto F(t, x, y(t, z))$$

has a minimum at x, therefore

$$0 = \partial_z F(t, x, y(t, z))|_{z=x}$$
$$= -(f^*)'\left(\frac{x - y(t, x)}{t}\right)\partial_x y(t, x) + u_0(y(t, x))\partial_x y(t, x).$$

Thanks to (11.11) and (11.22), we get

$$\partial_x w(t, x) = (f^*)'\left(\frac{x - y(t, x)}{t}\right) = g\left(\frac{x - y(t, x)}{t}\right) = u(t, x).$$

According to (iii), we have to prove that

$$\partial_x w \text{ is the unique entropy solution of (11.1)}.$$
(11.23)

We begin by proving that w solves

$$\begin{cases} \partial_t w + f(\partial_x w) = 0, & t > 0, \ x \in \mathbb{R}, \\ w(0, x) = \int_0^x u_0(s)ds, & x \in \mathbb{R}. \end{cases} \qquad (11.24)$$

We can first show that for every $t > 0$, $w(t, \cdot)$ is Lipschitz continuous. Fix $x, x' \in \mathbb{R}$, we have

$$\begin{aligned} w(t, x) - w(t, x') &= \min_{y \in \mathbb{R}} F(t, x, y) - F(t, x', y(t, x')) \\ &\le F(t, x, y(t, x') + x - x') - F(t, x', y(t, x')) \\ &= tf^* \left(\frac{x' - y(t, x')}{t} \right) + \int_0^{y(t,x')+x-x'} u_0(s)ds \\ &\quad - tf^* \left(\frac{x' - y(t, x')}{t} \right) - \int_0^{y(t,x')} u_0(s)ds \\ &= \int_{y(t,x')}^{y(t,x')+x-x'} u_0(s)ds \\ &\le \|u_0\|_{L^\infty(\mathbb{R})} |x - x'|, \end{aligned}$$

and since we can change the role of x and x' we gain

$$|w(t, x) - w(t, x')| \le \|u_0\|_{L^\infty(\mathbb{R})} |x - x'|. \qquad (11.25)$$

We prove now that

$$w(0, x) = \int_0^x u_0(s)ds, \qquad x \in \mathbb{R}. \qquad (11.26)$$

For every $t > 0$ and $x \in \mathbb{R}$

$$w(t, x) = \min_{y \in \mathbb{R}} F(t, x, y) \le F(t, x, x) = tf^*(0) + \int_0^x u_0(s)ds,$$

and

$$\begin{aligned} w(t, x) &= \min_{y \in \mathbb{R}} \left(tf^* \left(\frac{x - y}{t} \right) + \int_0^y u_0(s)ds \right) \\ &\ge \int_0^x u_0(s)ds + \min_{y \in \mathbb{R}} \left(tf^* \left(\frac{x - y}{t} \right) - \|u_0\|_{L^\infty(\mathbb{R})} |x - y| \right) \end{aligned}$$

$$= \int_0^x u_0(s)ds - t \max_{z \in \mathbb{R}} \left(\|u_0\|_{L^\infty(\mathbb{R})} |z| - f^*(z) \right)$$

$$= \int_0^x u_0(s)ds - t \max_{|v| \le \|u_0\|_{L^\infty(\mathbb{R})}} \max_{z \in \mathbb{R}} \left(vz - f^*(z) \right)$$

$$= \int_0^x u_0(s)ds - t \max_{|v| \le \|u_0\|_{L^\infty(\mathbb{R})}} f(v).$$

Then

$$\left| w(t, x) - \int_0^x u_0(s)ds \right| \le t \max \left\{ \max_{|v| \le \|u_0\|_{L^\infty(\mathbb{R})}} f(v), f^*(0) \right\},$$

as $t \to 0$ we get (11.26).

We show now that for every x the function $w(\cdot, x)$ is Lipschitz continuous. We preliminary prove the identity

$$w(t, x) = \min_{y \in \mathbb{R}} \left((t - \tau) f^* \left(\frac{x - y}{t - \tau} \right) + w(\tau, y) \right), \qquad 0 < \tau < t, \ x \in \mathbb{R}.$$
$$(11.27)$$

Fix $0 < \tau < t$ and $x \in \mathbb{R}$. Since f^* is convex and

$$\frac{x - y(\tau, x)}{t} = \left(1 - \frac{\tau}{t} \right) \frac{x - y}{t - \tau} + \frac{\tau}{t} \frac{y - y(\tau, x)}{\tau},$$

we have

$$f^* \left(\frac{x - y(\tau, x)}{t} \right) \le \left(1 - \frac{\tau}{t} \right) f^* \left(\frac{x - y}{t - \tau} \right) + \frac{\tau}{t} f^* \left(\frac{y - y(\tau, x)}{\tau} \right).$$

Then

$$w(t, x) = \min_{y \in \mathbb{R}} F(t, x, y) \le F(t, x, y(\tau, x))$$

$$= tf^* \left(\frac{x - y(\tau, x)}{t} \right) + \int_0^{y(\tau, x)} u_0(s)ds$$

$$\le (t - \tau) f^* \left(\frac{x - y}{t - \tau} \right) + \tau f^* \left(\frac{y - y(\tau, x)}{\tau} \right) + \int_0^{y(\tau, x)} u_0(s)ds$$

$$= (t - \tau) f^* \left(\frac{x - y}{t - \tau} \right) + w(\tau, y),$$

that gives

$$w(t, x) \leq \min_{y \in \mathbb{R}} \left((t - \tau) f^* \left(\frac{x - y}{t - \tau} \right) + w(\tau, y) \right).$$ (11.28)

On the other hand, writing

$$z = \frac{\tau}{t} x + \left(1 - \frac{\tau}{t} \right) y(t, x),$$

we have

$$\frac{x - z}{t - \tau} = \frac{x - y(t, x)}{t} = \frac{z - y(t, x)}{\tau}.$$

Then

$$(t - \tau) f^* \left(\frac{x - z}{t - \tau} \right) + w(\tau, z)$$

$$\leq (t - \tau) f^* \left(\frac{x - z}{t - \tau} \right) + \tau f^* \left(\frac{z - y(t, x)}{\tau} \right) + \int_0^{y(t,x)} u_0(s) ds$$

$$= t f^* \left(\frac{x - y(t, x)}{t} \right) + \int_0^{y(t,x)} u_0(s) ds = w(t, x),$$

that gives

$$w(t, x) \geq \min_{y \in \mathbb{R}} \left((t - \tau) f^* \left(\frac{x - y}{t - \tau} \right) + w(\tau, y) \right).$$ (11.29)

Clearly, (11.27) follows from (11.28) and (11.29).

Thanks to (11.25),

$$w(t, x) = \min_{y \in \mathbb{R}} \left((t - \tau) f^* \left(\frac{x - y}{t - \tau} \right) + w(\tau, y) \right)$$

$$\geq w(\tau, x) + \min_{y \in \mathbb{R}} \left((t - \tau) f^* \left(\frac{x - y}{t - \tau} \right) - \|u_0\|_{L^\infty(\mathbb{R})} |x - y| \right)$$

$$= w(\tau, x) - (t - \tau) \max_{z \in \mathbb{R}} \left(\|u_0\|_{L^\infty(\mathbb{R})} |z| - f^*(z) \right)$$

$$= w(\tau, x) - (t - \tau) \max_{z \in \mathbb{R}} \max_{|v| \leq \|u_0\|_{L^\infty(\mathbb{R})}} \left(vz - f^*(z) \right)$$

$$= w(\tau, x) - (t - \tau) \max_{|v| \leq \|u_0\|_{L^\infty(\mathbb{R})}} f(v),$$

and since we can change the role of t and τ we gain

$$|w(t, x) - w(\tau, x)| \le |t - \tau| \max_{|v| \le \|u_0\|_{L^\infty(\mathbb{R})}} f(v). \qquad (11.30)$$

Due to (11.25) and (11.30), w is Lipschitz continuous in both the variables and then it is differentiable almost everywhere in $(0, \infty) \times \mathbb{R}$. Let (t, x) be a point in which w is differentiable and $h > 0$, $\xi \in \mathbb{R}$. Thanks to (11.27)

$$w(t + h, x + h\xi) = \min_{y \in \mathbb{R}} \left(hf^* \left(\frac{x + h\xi - y}{h} \right) + w(t, y) \right)$$

$$\le hf^*(\xi) + w(t, x),$$

that is

$$\frac{w(t + h, x + h\xi) - w(t, x)}{h} \le f^*(\xi).$$

As $h \to 0$, we get

$$\partial_t w(t, x) + \xi \partial_x w(t, x) \le f^*(\xi), \qquad \xi \in \mathbb{R}.$$

Therefore

$$\partial_t w(t, x) + f(\partial_x w(t, x))$$
$$= \partial_t w(t, x) + \max_{\xi \in \mathbb{R}} (\xi \partial_x w(t, x) - f^*(\xi)) \le 0. \qquad (11.31)$$

On the other hand, for every $h > 0$, writing

$$\tau = t - h, \qquad z = \frac{\tau}{t} x + \left(1 - \frac{\tau}{t} \right) y(t, x),$$

we have

$$\frac{x - y(t, x)}{t} = \frac{z - y(t, x)}{\tau},$$

and

$$w(t, x) - w(\tau, z) \ge F(t, x, y(t, x)) - F(\tau, z, y(t, x))$$
$$= (t - \tau) f^* \left(\frac{x - y(t, x)}{t} \right).$$

Thus, for $h = t - \tau$,

$$\frac{w(t, x) - w\left(t - h, \left(1 - \frac{h}{t}\right)x + \frac{h}{t}y(t, x)\right)}{h} \geq f^*\left(\frac{x - y(t, x)}{t}\right).$$

As $h \to 0$,

$$\partial_t w(t, x) + \frac{x - y(t, x)}{t}\partial_x w(t, x) \geq f^*\left(\frac{x - y(t, x)}{t}\right),$$

and then

$$\partial_t w(t, x) + f(\partial_x w(t, x))$$
$$= \partial_t w(t, x) + \max_{\xi \in \mathbb{R}}(\xi \partial_x w(t, x) - f^*(\xi))$$
$$\geq \partial_t w(t, x) + \frac{x - y(t, x)}{t}\partial_x w(t, x) - f^*\left(\frac{x - y(t, x)}{t}\right) \geq 0. \qquad (11.32)$$

Equations (11.26), (11.31), and (11.32) say that w solves (11.24).

We claim that $u = \partial_x w$ is a weak solution of (11.1). Let $\varphi \in C^\infty(\mathbb{R}^2)$ be a test function with compact support. We have

$$0 = \int_0^\infty \int_{\mathbb{R}} (\partial_t w + f(\partial_x w))\partial_x \varphi \, dt dx$$
$$= -\int_0^\infty \int_{\mathbb{R}} w \partial_{tx}^2 \varphi \, dt dx - \int_{\mathbb{R}} w(0, x)\partial_x \varphi(0, x)dx + \int_0^\infty \int_{\mathbb{R}} f(u)\partial_x \varphi \, dt dx$$
$$= \int_0^\infty \int_{\mathbb{R}} (\partial_x w \partial_t \varphi + f(u)\partial_x \varphi)dt dx + \int_{\mathbb{R}} \partial_x w(0, x)\varphi(0, x)dx$$
$$= \int_0^\infty \int_{\mathbb{R}} (u \partial_t \varphi + f(u)\partial_x \varphi)dt dx + \int_{\mathbb{R}} u_0(x)\varphi(0, x)dx.$$

We conclude by proving that u satisfies an Oleinik type estimate. Since f is strictly convex there exists $c > 0$ such that

$$f'' \geq c > 0,$$

then, in light of Theorem 11.2,

$$0 \leq g' \leq \frac{1}{c}.$$

For every $t > 0$, $x \in \mathbb{R}$ and $h > 0$ we get

$$u(t, x) = g\left(\frac{x - y(t, x)}{t}\right) \geq g\left(\frac{x - y(t, x + h)}{t}\right)$$

$$\geq g\left(\frac{x + h - y(t, x + h)}{t}\right) - \frac{1}{c}\frac{h}{t} = u(t, x + h) - \frac{1}{c}\frac{h}{t},$$

that is

$$u(t, x + h) - u(t, x) \leq \frac{1}{c}\frac{h}{t}.$$

Since u is a weak solution of (11.1) satisfying an Oleinik type estimate, it is the unique entropy solution of (11.1) (see Chap. 10).   □

11.3   Exercises

Exercise 11.1 Let $a > 0$ be given. Prove that the Legendre transform of the function

$$f(u) = \frac{au^2}{2}$$

is

$$f^*(\xi) = \frac{\xi^2}{2a}.$$

Exercise 11.2 Let $n \in \mathbb{N} \setminus \{0\}$ be even. Prove that the Legendre transform of the function

$$f(u) = \frac{u^n}{n}$$

is

$$f^*(\xi) = \frac{\xi^m}{m}, \qquad \text{with } m = \frac{n}{n - 1}.$$

Exercise 11.3 The Legendre transform of non strictly convex functions is not well-defined. Let $a \in \mathbb{R}$ be given. Prove that the Legendre transform of the function

$$f(u) = au$$

is

$$f^*(\xi) = \begin{cases} \infty, & \text{if } \xi \neq a, \\ 0, & \text{if } \xi = a. \end{cases}$$

Exercise 11.4 Let f and g satisfy (11.5), $a \in \mathbb{R}$ and $\lambda \neq 0$ be given. Prove the following identities[1]

$$(\lambda f)^*(\xi) = \lambda f^* \left(\frac{\xi}{\lambda} \right), \tag{11.33}$$

$$(f(\lambda \cdot))^*(\xi) = f^* \left(\frac{\xi}{\lambda} \right), \tag{11.34}$$

$$(f + a)^*(\xi) = f^*(\xi) - a, \tag{11.35}$$

$$(f(\cdot - a))^*(\xi) = f^*(\xi) + a\xi, \tag{11.36}$$

$$f \leq g \Rightarrow g^* \leq f^*, \tag{11.37}$$

$$\inf_{u \in \mathbb{R}} (f(u) + g(u)) = \sup_{\xi \in \mathbb{R}} (-f^*(-\xi) - g^*(\xi)). \tag{11.38}$$

References

1. Brezis, H.: Functional Analysis, Sobolev Spaces and Partial Differential Equations. Universitext. Springer, New York (2011)
2. Karlsen, K.H., Risebro, N.H.: A note on front tracking and the equivalence between viscosity solutions of Hamilton-Jacobi equations and entropy solutions of scalar conservation laws. Nonlinear Anal. Theory Methods Appl. Ser. A Theory Methods **50**(4), 455–469 (2002)
3. Lax, P.D.: Hyperbolic systems of conservation laws. II. Commun. Pure Appl. Math. **10**, 537–566 (1957)
4. Oleinik, O.A.: On Cauchy's problem for nonlinear equations in a class of discontinuous functions. Doklady Akad. Nauk SSSR (N.S.) **95**, 451–454 (1954)

---

[1] Equation (11.38) is known as the Fenchel-Rockafellar duality formula [1, Theorem 1.12].

Index

A
Absolutely continuous functions, 60
Agmon Theorem, 100
Asymptotic behavior, 116
Aw-Rascle model, 4

B
Bounded solutions, 108
Burgers equation, 4
BV, 59

C
Characteristics, 7
Compensated compactness, 103
Conserved quantity, vii, 1

D
Div-Curl Lemma, 101

E
Entropy, 24
Entropy flux, 24
Entropy solution, 25
Error estimate, 88
Euler equations, 4

F
Fenchel-Rockafellar duality formula, 145
Flux, vii, 1
Front-tracking, 73

G
Genuine nonlinearity, 95

H
Hamilton-Jacobi equation, 131
Heaviside function, 60
Helly Theorem, 67

K
Kružkov entropies, 26
Kružkov Theorem, 37

L
Lax–Oleinik formula, 135
Legendre transform, 132
Lighthill-Whitham-Richards model, 3

M
Murat Lemma, 100

O
Oleinik estimate, 122

P
Periodic solutions, 113
Piecewise constant functions, 60

Q
Quasilinear equation, 11